Springer-Verlag Berlin Heidelberg GmbH

31 Fortschritte der chemischen Forschung
Topics in Current Chemistry

Stereo- and Theoretical Chemistry

Springer-Verlag Berlin Heidelberg GmbH 1972

ISBN 978-3-540-05841-0   ISBN 978-3-540-37476-3 (eBook)
DOI 10.1007/978-3-540-37476-3

Contents

Stereochemical Reaction Cycles [1)]

Prof. Donald J. Cram and Dr. Jane Maxwell Cram

University of California, Department of Chemistry, Los Angeles, USA

Contents

1. Introduction

The concept and first examples of stereochemical reaction cycles were introduced by Walden in connection with his discovery of the Walden or optical inversion [2]. The cycle of Phillips and Kenyon [3] (Chart I) historically demonstrated that the reaction of bimolecular substitution at carbon proceeded with inversion of configuration. Chart I illustrates the generally recognized principle that when a cycle contains an odd number of reactions that go with inversion of configuration, two enantiomers must be included in the cycle.

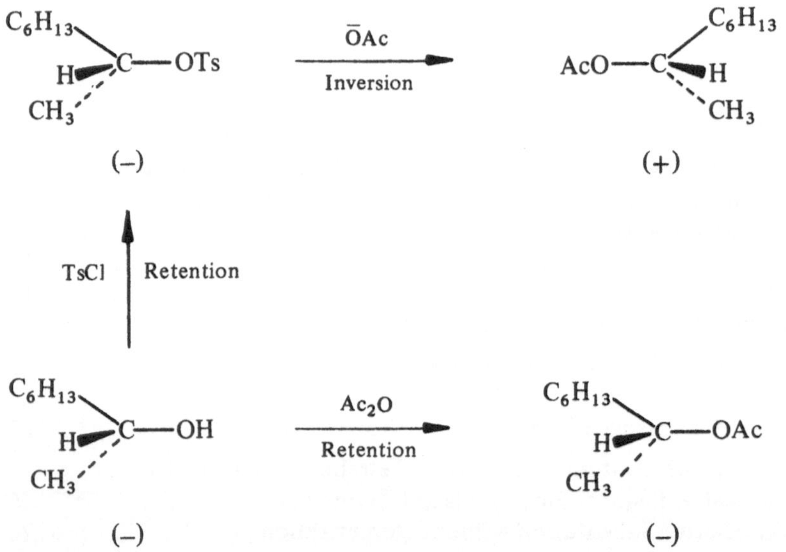

Chart I

The more recently discovered *stereochemical reaction cycle* [4] of Chart II presents an exception to this "conventional wisdom". The cycle contains one reaction that occurs with inversion of configuration, and two that go with retention. However, two enantiomerically related compounds are not found in this cycle. This exception can not be dismissed as *"something of a stereochemical curiosity"* [5], particularly since other departures from the rule are in the literature [5-7].

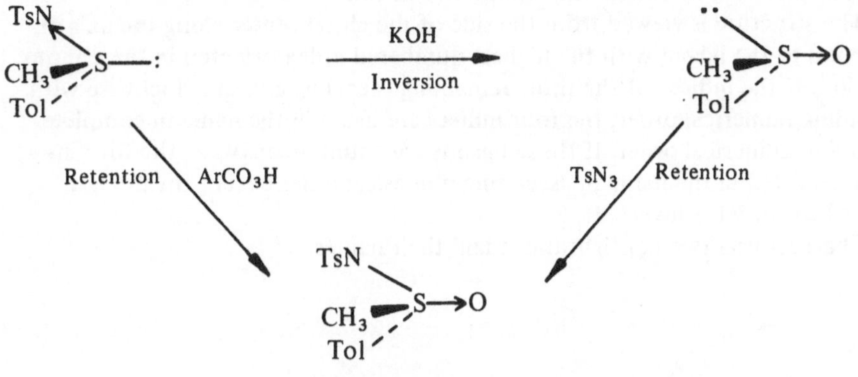

Chart II

In this paper, the principles and properties of stereochemical reaction cycles are discussed. Maps are developed that provide a means for monitoring configurational changes. As cycles become more complex, configurational bookkeeping becomes more important. The coupled tasks of correlation and prediction are made simpler and more interesting by maps and by the generalizations that make them useful. Unfortunately, new terms and conventions are a necessary encumbrance.

We are concerned here with several types of stereochemical reaction cycles. Cycles that involve simple substitution reactions on tetrahedra are discussed in the second section. Cycles composed of simple substituent exchange reactions of atropisomers occupy the third section. In the fourth we treat the effect on cycles of associative substitution reactions of tetrahedra that involve rearrangeable trigonal bipyramids as intermediates.

2. Stereochemical Cycles Composed of Simple Substitution Reactions on Tetrahedra [1]

2.1 Characterization

Chiral compounds with four different ligands attached tetrahedrally to a chiral center are represented by four different numerical indices or digits, one for each ligand. Electron pairs that occupy an apex of a tetrahedron are treated as ligands. The ligands that do not undergo exchange (static ligands) are given the lower numerical indices, which are listed first in the name in ascending numerical order. The ligands that undergo exchange are given the higher in-

3

dices, and their order specifies the configuration, particularly the last two of the four indices. *The convention adopted* is as follows:

The structure is viewed from the side of the chiral center along the axis of the bond to the ligand with the highest numbered index oriented in the remote position. If the indices of the three remaining (near) ligands are clockwise with ascending numerical order, the four indices are listed in the name in complete ascending numerical order. If those ligands are counterclockwise, the four indices are listed in the name in ascending numerical order except for the last two, whose order is inverted.

Chart III lists two configurations and their names.

1234   1243

Chart III

Complications of reaction mechanism are avoided by an operational definition of a reaction. A reaction converts a compound of specifiable structure into a second of specifiable structure in a stereospecific process. The new compound possesses one new ligand not present in the starting material. For example,

$$1234 \longrightarrow 1235$$

means ligand 4 was replaced or modified to give the new ligand 5. When two ligands are modified in the same reaction, the reaction is represented as two reactions occurring in sequence. Thus reaction

$$1234 \longrightarrow 1256$$

is formally composed of

$$1234 \longrightarrow 1236 \longrightarrow 1256.$$

There are two possible stereochemical courses for a ligand substitution reaction at a tetrahedral chiral center, inversion or retention of configuration. The terms inversion and retention apply only to the outcome of individual reactions, and are not used in reference to the "outcome" of a stereochemical cycle. The letters R (retention) or I (inversion) can be placed over reaction arrows to emphasize the stereochemical course of a reaction although the same information is deducible by comparison of the sequences of indices of starting material and product.

$$1234 \xrightarrow[-4,\,+5]{R} 1235$$

$$1234 \xrightarrow[-4,\,+5]{I} 1253$$

Reaction chains are sequences of transformations in which the product of one reaction serves as starting material for the next. A *cycle* is a closed chain of ligand substitution reactions that connects compounds in a circular sequence In this use of the word compound, enantiomers are not distinguished. In a reaction cycle, the number of reactions equals the number of compounds. The directions of the arrows for the reactions of a cycle are incidental to the properties of the cycle. In the generalized three-reaction, three-compound cycle formulated, any of the arrows could point in either direction.

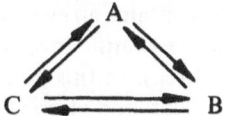

A *stereochemical* reaction cycle is one in which all compounds are chiral and all reactions stereospecific. Although the number of compounds equals the number of reactions in a cycle, the number of *chiromers* may either be equal to or exceed the number of reactions by one. The term *chiromer* denotes any member of a set of optically pure compounds related to one another by their inclusion in the same stereochemical reaction cycle. Each enantiomer of one chiral compound may be included as a separate chiromer in a stereochemical reaction cycle. Inclusion of two sets of enantiomers provides reducible cycles that are separable into two simpler reaction cycles.

When the number of chiromers equals the number of reactions, the cycle is termed *podal.* when the number of chiromers exceeds the number of reactions by one the cycle is *antipodal.* As depicted, the podal cycle of Chart IV starts and ends on the same chiromer. The antipodal cycle begins and ends on enantiomerically related chiromers.

5

$$1234 \xrightarrow{\quad R \quad} 1235 \xrightarrow{\quad R \quad} 1236 \xrightarrow{\quad R \quad} 1234$$

podal stereochemical reaction cycle
(chiromers = reactions = three)

$$1234 \xrightarrow{\quad R \quad} 1235 \xrightarrow{\quad R \quad} 1236 \xrightarrow{\quad I \quad} 1243$$

antipodal stereochemical reaction cycle
(chiromers = reactions + 1 = four)

Chart IV

Reaction cycles are further distinguished by the number of static ligands common to each chiromer. In both of the cycles of Chart IV, ligands 1, 2 and 3 are common to each chiromer, and ligands 4, 5 and 6 change. A reaction cycle in which three ligands are static is termed a *triligostat*.

Triligostatic cycles are the kind most often encountered in the literature. Their simplicity produced the classical rule that *an even number of inversions in a reaction chain gives product of overall retained configuration, whereas an odd number gives product of the enantiomeric configuration.*

Apparent violations of this rule are encountered in *diligostatic cycles* (two static ligands common to all chiromers), in *monoligostatic* (one static ligand) and in *aligostatic cycles* (no static ligands).

The simplest cycles that violate this rule are diligostatic. Chart V provides an example of an antipodal cycle that contains an even number (zero) of inversions for individual reactions. The second example involves an odd number of inversions for individual reactions (three). In this podal cycle, the starting chiromer is regenerated in the product. Thus the classical rule is violated from two directions.

$$1234 \xrightarrow{\quad R \quad} 1235 \xrightarrow{\quad R \quad} 1245 \xrightarrow{\quad R \quad} 1243$$

diligostatic antipodal three-reaction stereochemical
cycle (chiromers = reactions plus 1)

$$1234 \xrightarrow{\quad I \quad} 1253 \xrightarrow{\quad I \quad} 1245 \xrightarrow{\quad I \quad} 1234$$

diligostatic podal three-reaction stereochemical
cycle (chiromers = reactions)

Chart V

A property common to the two cycles of Chart V is that, in the first reaction, ligand 5 replaces 4. In the second, ligand 4 replaces 3. In the third, ligand

3 replaces 5. This overall interchange of ligands and bonds is independent of the stereochemical course (R or I) of the particular reactions that participate in the interchange. In both reaction cycles, ligands 3 and 4 are both replaced at one stage or another, and therefore ligands 3 or 4 are absent from certain chiromers. When a pair of ligands interchange their bonding sites on a chiral tetrahedron by a series of ligand substitution reactions (each ligand leaves and returns), the process is called a *ligand metathesis,* or an *LM.* Clearly, an *LM* must be a property of a reaction cycle rather than of a particular reaction. The smallest cycle that can contain an *LM* involves three reactions.

The stereochemical cycles of Chart V each contain an *LM.* Inclusion of an *LM* in a cycle has the same effect on the number of chiromers of the cycle as does the reversal of the stereochemical course of any single reaction. For example, if a cycle contains two I's and no *LM,* the cycle is podal. If a cycle contains an *LM* and one I, it is also podal. But cycles with two I's and an *LM* or with three I's and no *LM* are antipodal.

The differences between an inversion (I) and a ligand metathesis (LM) require emphasis.

1) An inversion is characteristic of a particular reaction, not of a series of reactions. A ligand metathesis is not characteristic of a particular reaction, but of a series of reactions.

2) An inversion involves substitution of one ligand for another as two bonds interchange their positions with respect to the other two bonds. A ligand metathesis involves interchange of two ligands and two bonds by a sequence of substitution reactions that involves each ligand leaving and reentering the molecule.

3) In an inversion, bonds must move. In a ligand metathesis, bonds may or may not move. If they move, track must be kept of that movement when bonds and ligands are paired.

Chart VI illustrates the differences between an inversion and a ligand metathesis. Bonds are labeled 1', 2' etc. in this scheme, and although they may move with respect to one another during a reaction, they always remain with the chiral center.

In the definition of a stereochemical reaction cycle, only one set of enantiomers can be included among the chiromers. This fact in itself limits the number of *LM*'s in a cycle to either zero or one, the former being an even, and the latter an odd number. More complex closed reaction chains that contain more than one set of enantiomers can be envisioned. These cycles always are reducible into two or more non-reducible reaction cycles. For example, the following cycle is reducible:

$$1234 \longrightarrow 1235 \longrightarrow 1236 \longrightarrow 1253 \longrightarrow 1243.$$

This reaction chain is composed of the nonreducible cycle, $1235 \longrightarrow 1236 \longrightarrow 1253$ with the reactions $1234 \longrightarrow 1235$ and

7

Reaction that goes with inversion
(bonds 3ʹ and 4ʹ exchange their positions with respect to the other two bonds)

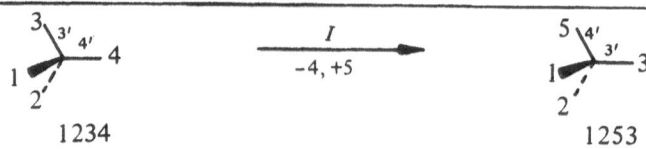

Cycle antipodal because of ligand metathesis

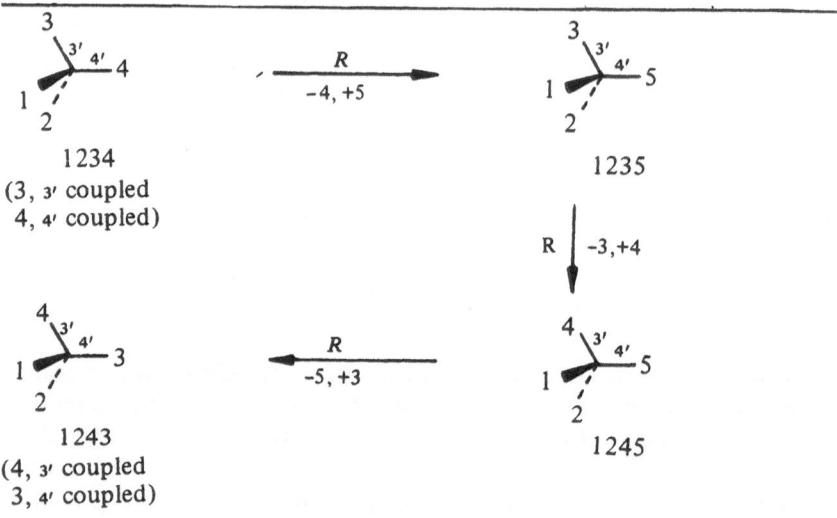

Cycle podal because of ligand metathesis

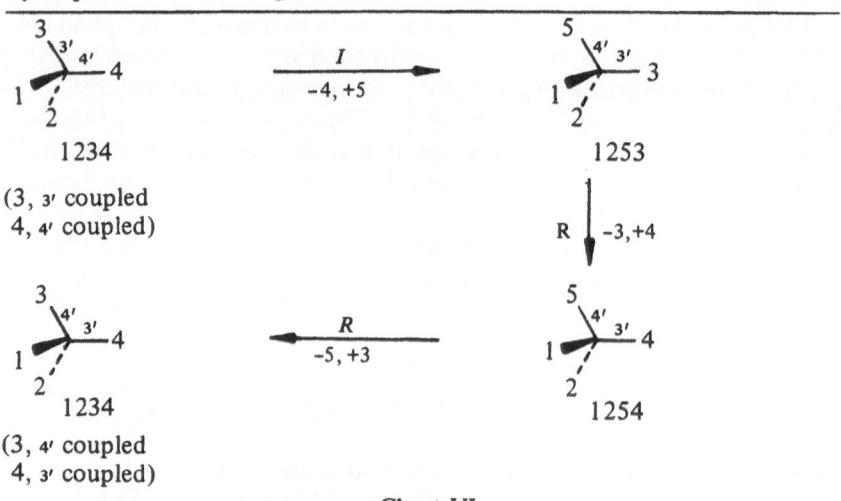

Chart VI

1253 $\longrightarrow$ 1243 left over. Reaction 1253 $\longrightarrow$ 1243 makes the reaction 1235 $\longrightarrow$ 1234 predictable. The combination of 1234 $\longrightarrow$ 1235 and predictable reaction 1235 $\longrightarrow$ 1234 provides the second full reaction cycle, 1234 $\longrightarrow$ 1235 $\longrightarrow$ 1234.

The above principles serve to establish the symmetry properties of a stereochemical reaction cycle of any length or type. *When the sum of the number of I's and LM's in a cycle is even, the cycle is podal. The cycle is antipodal if the sum is odd.*

Podal cycles with an even number of *I*'s and antipodal cycles with an odd number of *I*'s conform to the classical rules, and do not contain *LM*'s. Podal cycles that contain an odd number of *I*'s and antipodal cycles that contain an even number of *I*'s must each contain an *LM*. These types of cycles were not taken into account in the classical concepts. A single exception is the work of E. Fischer and F. Brauns [6], who carried out the reactions of Chart VII for the purpose of illustrating a ligand metathesis. In this cycle, no bonds to the chiral

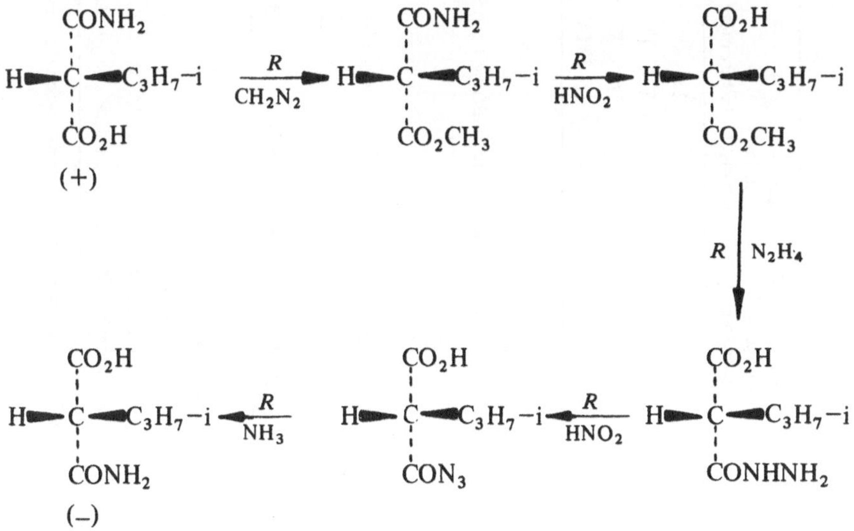

Diligostatic, antipodal, six-reaction stereochemical cycle that contains one *LM*

Chart VII

center are made or broken, and therefore the reactions all go with retention of configuration. Since two ligands (H and i-C_3H_7) are common to all chiromers, the cycle is diligostatic. The cycle contains five reactions and six chiromers, is antipodal, and contains one *LM*.

9

Table I. *General properties of stereochemical reaction cycles composed of tetrahedral chiromers of cycle size X[a]*

Cycle type	Minimum cycle size[a]	Substitutable ligand positions	Substitutable ligands[b]	Maximum number	
				Chiromers	Allowed reactions[c]
Triligostat	2	1	X	$2X$	$2X(X-1)$
Diligostat	3	2	X	$X(X-1)$	$2X(X-1)(X-2)$
Monoligostat	4	3	X	$1/3X(X-1)(X-2)$	$X(X-1)(X-2)(X-3)$
Aligostat	5	4	X	$1/12X(X-1)(X-2)(X-3)$	$1/3X(X-1)(X-2)(X-3)(X-4)$

[a]) X = number of reactions of a cycle.
[b]) All ligands involved in substitution.
[c]) Direct enantiomer interconversions and simultaneous multiple ligand substitutions are excluded.

Table II. *Generalizations about presence or absence of a ligand metathesis (LM) in cycles of various sizes and types*

Cycle size	Triligostat	Diligostat	Monoligostat	Aligostat
2	Zero LM	a	a	a
3	Zero LM	One LM	a	a
4	Zero LM	Zero or one LM	Zero LM	a
5	Zero LM	Zero or one LM	Zero or one LM	One LM
6	Zero LM	Zero or one LM	Zero or one LM	Zero or one LM

a = The number of reactions is below the minimum required for the reaction cycle

2.2 Properties

Above, stereochemical reaction cycles that involve tetrahedral chiral centers are characterized according to the number of static ligands. Each of the four types (triligostat, diligostat, monoligostat and aligostat) has a set of properties which vary with the *cycle size* as measured by the *number of reactions*. Each type of cycle has a minimum size, and cannot exist with fewer reactions. Each type of cycle has a fixed number of substitutable ligand positions. Each cycle of a given size and type has a maximum number of replaceable ligands that enter or leave the chiromers during the reactions of the cycle. Finally, each cycle of a given size and type has a maximum number of allowed reactions usable for connecting the different chiromers. Certain reactions are forbidden. A direct interconversion of enantiomers is excluded because the reaction involves a stereospecific substitution of one ligand by an identical ligand. A stereochemical cycle of only one compound and one reaction does not exist in practice. The entering and leaving ligands are indistinguishable, and racemization results. Another forbidden reaction is the simultaneous substitution of more than one ligand. Such reactions are factored into a sequence of single ligand substitutions. Table I lists the general properties which have been derived [1] for cycles of given sizes and types.

A property (not listed in Table I) of cycles of various sizes and types is the possible presence or absence of a ligand metathesis (*LM*). Table II summarizes the generalizations concerning this property. For triligostats of any size, no *LM* is possible. The three-reaction diligostats and five-reaction aligostats each *must contain* an *LM*. The four-reaction monoligostat can be closed only *without* an *LM*. The remaining cycles either can contain zero or one *LM*, depending on which reactions are used.

2.3 Maps

The alternative reaction pathways for each cycle size and shape are best visualized with maps. The maps are composed of symbols for all possible chiromers appropriately connected with one another by lines that represent permitted reactions. Simple polyhedra provide maps that correlate all two- and three-reaction cycles (Figs. 1 and 2). In these polyhedra the apices represent all possible chiromers, and the edges all the allowed reactions. Four-reaction cycles are mapped with either cubes (Figs. 3 and 4) or a planar graph (Fig. 5). Fig. 6 is a similar map for the five-reaction aligostatic cycles. In the planar graphs of Fig. 5 and 6, circles represent the chiromers and the straight connecting lines the reactions. The fact that many of the connecting lines of Figs. 3 – 6 intersect has no significance. The numbers of chiromers and reactions for each of these maps were derived from the generalizations of Table I.

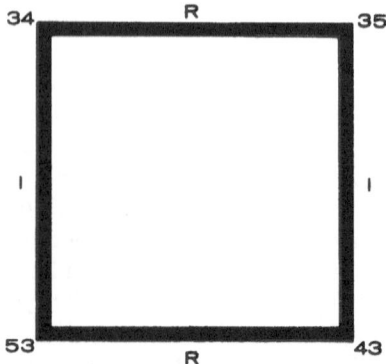

Fig. 1. Map of two-reaction stereochemical cycle

In these maps, the chiromers are identified by the indices that represent their ligands, and the configuration is specified by the order of the indices according to the convention developed above. In the maps of Fig. 1, 2, 3 and 5, only the last two of the total of four digits are listed since digits 12 are common to all chiromers and are understood. In the map of Fig. 3, the last three of the four indices are listed since index 1 is common to all chiromers and is understood. In the map of Fig. 6, all four indices require listing. The symbols R (retention) or I (inversion) associated with each of the connecting lines (reactions) specify the stereochemical course of each reaction. In the discussion of these reaction cycles, the same conventions apply as apply to the maps. Only those indices will be used that are needed to specify ligand changes and configurations.

In these maps, enantiomers always are found opposite one another. The two-reaction cycle must be antipodal and triligostatic (Fig. 1). The three-reaction cycles (Fig. 2) either may be of the triligostatic or diligostatic type. In the square-pyramidal map of the triligostatic type, the three reactions denoted by the edges of any face of the polyhedron are podal. These reaction triangles involve either three retentions (RRR), or one retention and two inversions (IIR). An example of the latter type is

$$35 \longrightarrow 43 \longrightarrow 36 \longrightarrow 35.$$

To close an antipodal three-reaction triligostatic cycle, the path must involve three edges, two of which are not common to the same triangle, and one of which is common to two triangles. These cycles all involve four chiromers and three reactions that go with inversion (III), or one inversion and two retentions (RRI). An example of the former type is

$$34 \longrightarrow 53 \longrightarrow 36 \longrightarrow 43.$$

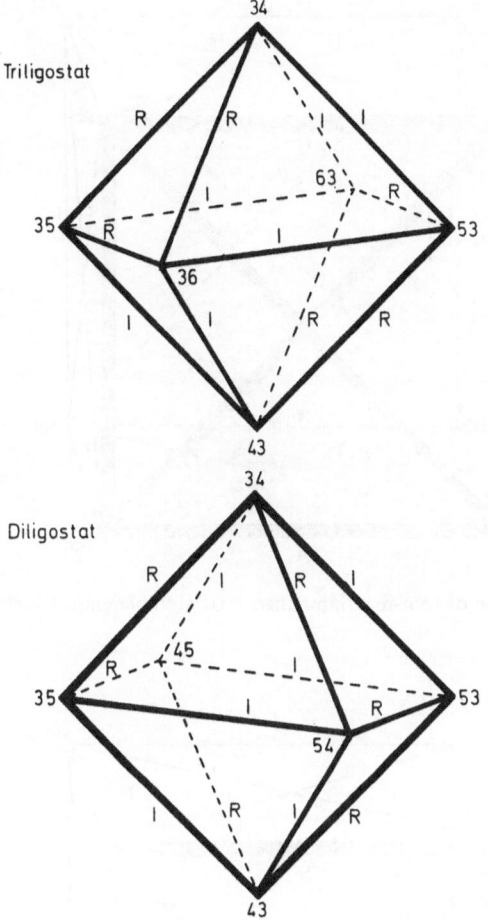

Fig. 2. Maps of three-reaction stereochemical cycles

In the square pyramidal map of the three-reaction diligostatic variety (Fig. 2), again podal cycles are found on the triangular faces. The three reactions may all go with inversion (*III*), or one inversion and two retentions (*IRR*). An example of *III* is

$$34 \longrightarrow 53 \longrightarrow 45 \longrightarrow 34.$$

An example of *IRR* is

$$34 \longrightarrow 53 \longrightarrow 54 \longrightarrow 34.$$

13

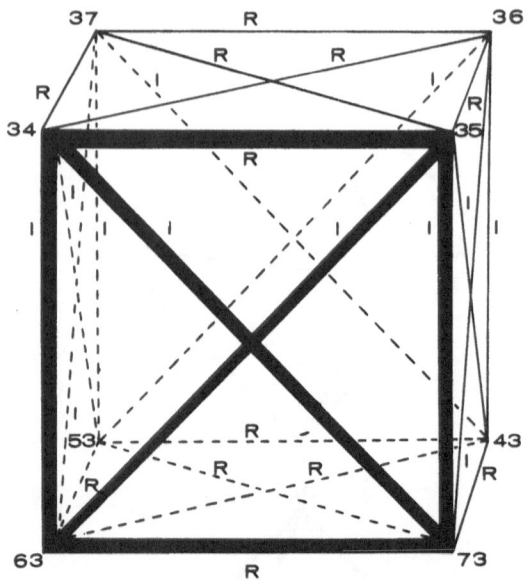

Fig. 3. Map of four-reaction triligostatic stereochemical cycle

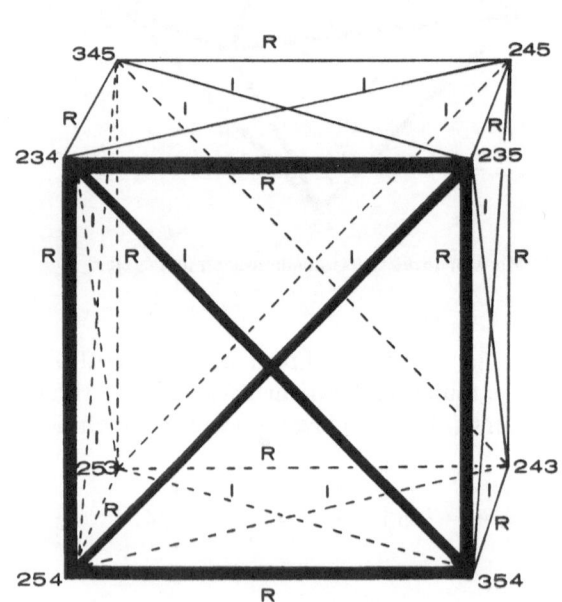

Fig. 4. Map of four-reaction monoligostatic stereochemical cycle

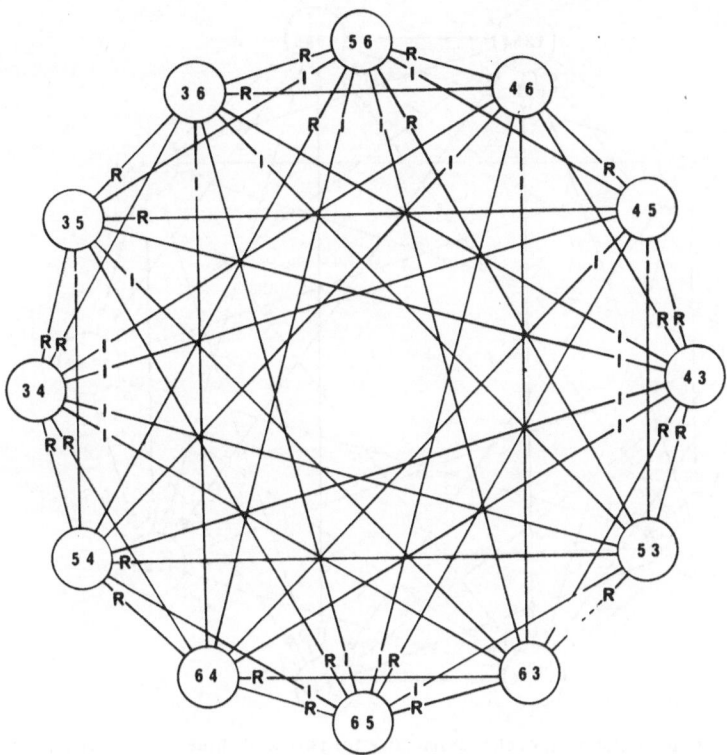

Fig. 5. Map for stereochemical cycles, exhaustive for two- and three-reaction triligostatic and for three- and four-reaction diligostatic types, and illustrative for five- and six-reaction diligostatic types

To close a three-reaction antipodal diligostatic cycle, the path must involve three edges, two of which are not common to the same triangle and one of which is common to two triangles. These cycles must involve either three retentions (*RRR*), or two inversions and one retention (*IIR*). An example of the former is 34 $\longrightarrow$ 35 $\longrightarrow$ 45 $\longrightarrow$ 43, and of the latter cycle is 34 $\longrightarrow$ 53 $\longrightarrow$ 45 $\longrightarrow$ 43.

All possible four-reaction cycles are mapped in Fig. 3, 4 and 5. Of these maps, the most complex is that of the diligostatic type, in which twelve chiromers are connected by forty-eight reactions (Table I). Generalizations that indicate what other cycles are included in an exhaustive representation of a cycle of given size and type have been stated elsewhere [1].

Five reactions are the smallest number needed to construct an aligostat (Fig. 6), and each possible cycle must contain an *LM*. An example of a podal cycle is

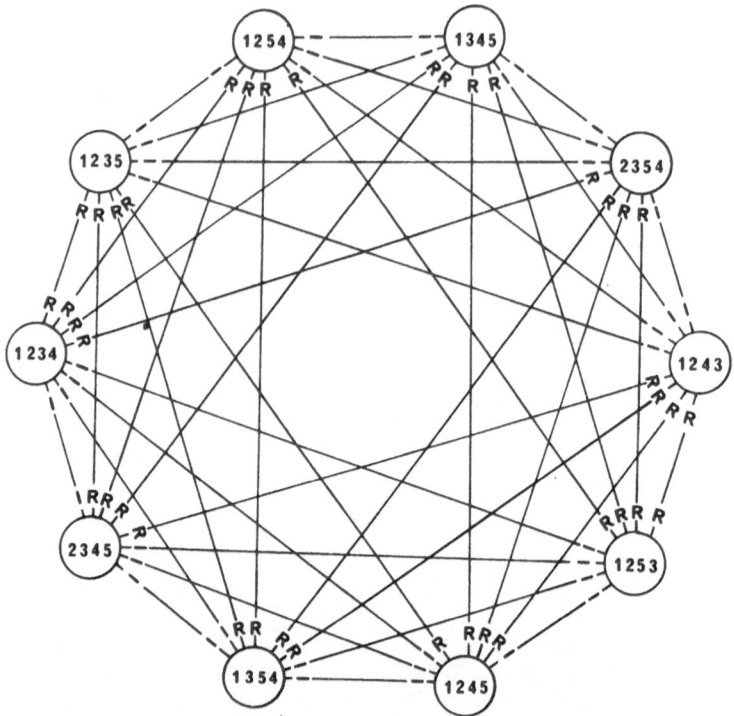

Fig. 6. Map for stereochemical cycles, exhaustive for two- and three-reaction triligostatic three-reaction diligostatic, four-reaction monoligostatic and five-reaction aligostatic types

$$1234 \xrightarrow{R} 1235 \xrightarrow{R} 1245 \xrightarrow{R} 1345 \xrightarrow{R} 2345 \xrightarrow{I} 1234$$

This cycle contains an *LM* and one *I*. An antipodal cycle that contains only an *LM* is

$$1234 \xrightarrow{R} 1235 \xrightarrow{R} 1245 \xrightarrow{R} 1345 \xrightarrow{R} 2345 \xrightarrow{R} 1243$$

2.4 Applications

Stereochemical reaction cycles sometimes aid in assignment of relative configurations to compounds and in determination of the stereochemical course of substitution reactions. For example, if mechanistic analogy suggests all reactions of a three-reaction cycle involve the same steric course, then that course (*R* or *I*) is determined alone by knowledge of whether three or four chiromers are included in the cyclic (podal or antipodal), and by whether the cycle is tri- or diligostatic. With any cycle, comparison of the number of reactions and chiromers and identification of the cycle type determines whether an even or odd number of individual reactions go with inversion. If other information is available concerning the steric course of some of the reactions, the possibilities are narrowed for the steric courses of the remaining reactions. If the steric courses of all but one of the reactions are known, the steric course of that one reaction is established by the cycle's properties, and all configurations are unambiguously related.

Chart I outlines the most quoted example of the *Walden inversion cycle* with carbon as the chiral center:

$$34 \quad \xleftarrow{\;R\;} \quad 35 \quad \xrightarrow{\;R\;} \quad 36 \quad \xrightarrow{\;I\;} \quad 43$$

(–)-Acetate	(–)-Alcohol	(–)-Tosylate	(+)-Acetate

This three-reaction stereochemical cycle is antipodal and triligostatic. A different type of cycle that involves carbon was closed by Bernstein and Whitmore [8] (Chart VIII). The cycle is large and is composed of many elegant conversions. This interesting cycle is podal, diligostatic, and contains one *I* and and one *LM*. Although the deaminative rearrangement is a simultaneous double ligand substitution, the character of the cycle is not affected, and can be counted as two reactions in the "bookkeeping".

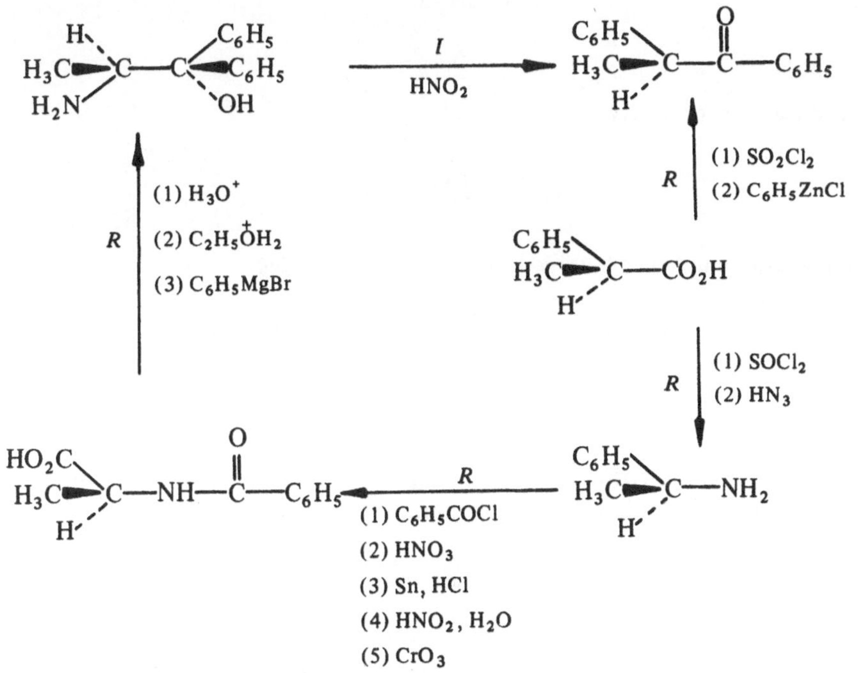

Chart VIII

17

Several examples of different kinds of stereochemical reaction cycles with chiral sulfur are available, some of which involve an electron pair as a ligand. Chart IX provides several examples [4].

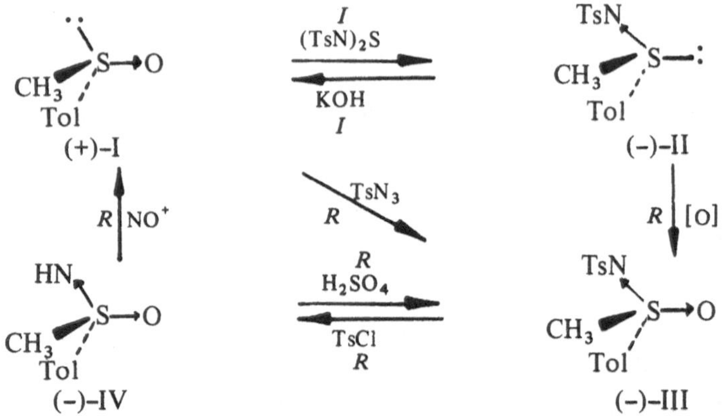

Tol= $p-CH_3C_6H_4$; Ts = $p-CH_3C_6H_4SO_2$

Chart IX

The two-reaction cycle

$$(+)\text{-}I \underset{I}{\overset{I}{\rightleftharpoons}} (-)\text{-}II$$

is triligostatic, podal and involves two I's. The two cycles

$$(-)\text{-}III \xleftarrow{R} (+)\text{-}I \xrightarrow{I} (-)\text{-}II \xrightarrow{R} (-)\text{-}III$$

$$(-)\text{-}III \xleftarrow{R} (-)\text{-}II \xrightarrow{I} (+)\text{-}I \xrightarrow{R} (-)\text{-}III$$

are three-reaction, diligostatic in type (CH_3 and Tol are static ligands), are podal (chiromers=reactions) and each contains one I and an LM. The cycle

$$(+)\text{-}I \xrightarrow{I} (-)\text{-}II \xrightarrow{R} (-)\text{-}III \xrightarrow{R} (-)\text{-}IV \xrightarrow{R} (+)\text{-}I$$

is four-reaction, diligostatic, podal and contains one I and an LM.

Chart X contains the only example [9] to the authors' knowledge of a mono-ligostatic stereochemical reaction cycle. In this cycle, only the tolyl group is common to all six chiromers, and the following ligands undergo exchange: 0, electron pair, NTs, NH, NCH$_3$ and TsNCH$_3$. The variety of ligand types and reactions for their substitution associated with sulfur as a chiral center is unparalleled except possibly in the chemistry of phosphorus. The only type of cycle that remains experimentally unexemplified is the aligostat.

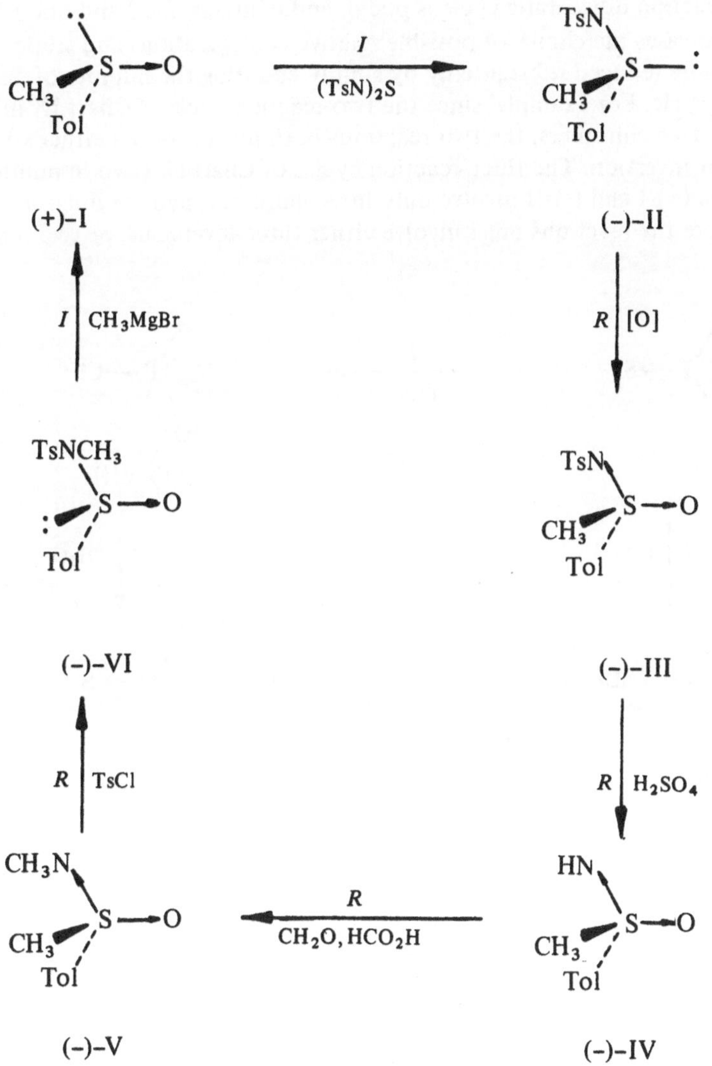

(+)-I

(−)-II

(−)-VI

(−)-III

(−)-V

(−)-IV

Tol = p−CH$_3$C$_6$H$_4$;   Ts = p−CH$_3$C$_6$H$_4$SO$_2$

Chart X

19

The six-reaction, monoligostatic cycle of Chart X is podal, contains two
I's, four R's and no LM's. By appropriately blending selected reactions of
Charts IX and X, another interesting cycle is composed:

$$(+)\text{-}I \xrightarrow{R} (-)\text{-}III \xrightarrow{R} (-)\text{-}IV \xrightarrow{R} (-)\text{-}V \xrightarrow{R} (-)\text{-}VI \xrightarrow{I} (+)\text{-}$$

This five-reaction diligostatic cycle is podal, and contains one I and one LM.

In some cases the choice of possible relative configurations and steric
courses can be reduced substantially by simply counting the number of chiromers in a cycle. For example, since the two-reaction cycle of Chart IX involves only two chiromers, the two reactions both must proceed either with
retention or inversion. The three-reaction cycles of Chart IX (two in number)
that include (+)-I and (-)-II involve only three chiromers and are diligostatic,
and therefore the reactions must involve either three-inversions, or two reten-

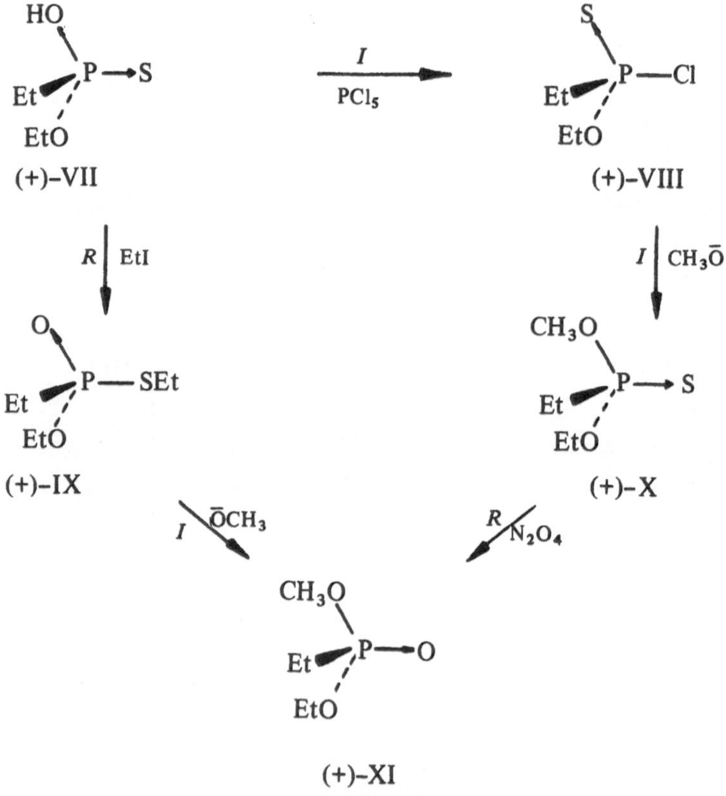

(+)-VII

(+)-VIII

(+)-IX

(+)-X

(+)-XI

Chart XI

tions and one inversion. The three-reaction cycles of Chart IX that are triligostatic involve three chiromers. Therefore the reactions either must all proceed with retention, or with one retention and two inversions.

Of the variety of stereochemical reaction cycles centered around tetrahedral phosphorus, the diligostatic types are the most interesting. The five-reaction, podal cycle of Chart XI [10)] contains three I's, two R's and one LM. An unusual feature of the cycle is the conversion of (+)-VII to (+)-IX. In this reaction, two ligands were changed: OH became O and S became SEt (Et and

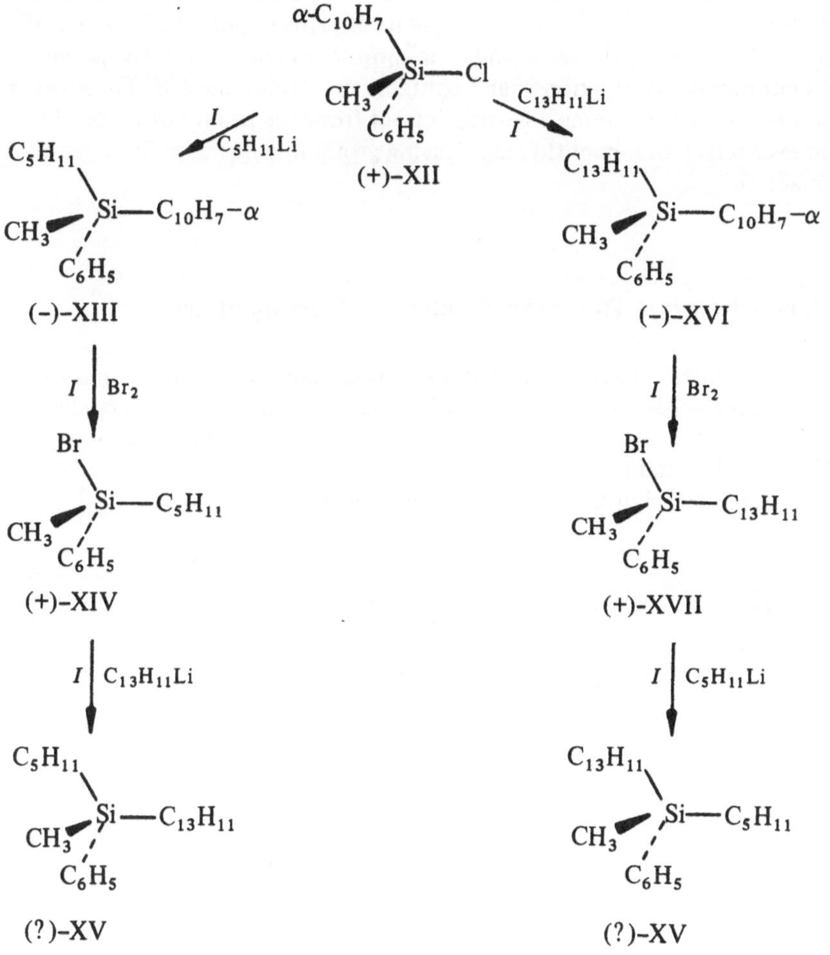

α-$C_{10}H_7$ = α-naphthyl; C_5H_{11} = neopentyl; $C_{13}H_{11}$ = benzhydryl

Chart XII

21

OEt were the static ligands). Although this double ligand reaction could be factored into two reactions (both going with retention), the system is simple enough to analyze without this complication.

Many elegant examples of stereochemical reaction cycles involving chiral silicon are known [11], all of which are triligostatic in type. One diligostatic cycle has been developed in principle, although not in practice [12]. Although all of the critical reactions formulated in Chart XII have been carried out and their steric course determined, reactions (+)-XIV $\longrightarrow$ (?)-XV and (+)-XVII $\longrightarrow$ (?)-XV have not been reported. However, that these reactions would occur if tried and would go with inversion has been demonstrated amply by the other reactions of the cycle itself. This hypothetical cycle is diligostatic (CH_3 and C_6H_5 are ligands common to all chiromers), antipodal (seven chiromers, six reactions), and contains six *I*'s and one *LM*. The absence of the two missing reactions does not detract from the beauty of cycle. Who would ever select an α-naphthyl as a leaving group in a nucleophilic substitution reaction?

3. Stereochemical Reaction Cycles of Atropisomers

Molecules chiral due to restricted rotation about single or double bonds (atropisomers) may act as the only chiromers in stereochemical reaction cycles. Each individual reaction is presumed to occur with retention, although conceivably a reaction could be induced to occur with inversion. Such a possibility is outlined in Chart XIII, but is contrived and specialized.

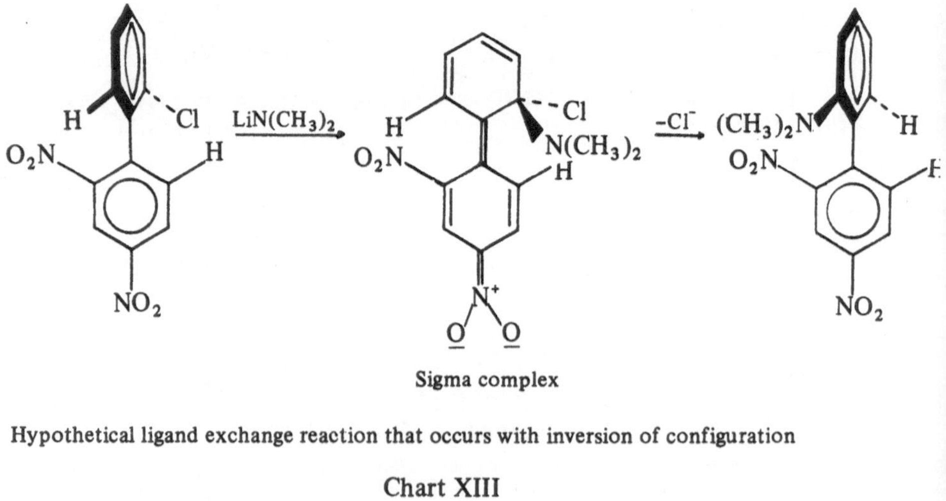

Sigma complex

Hypothetical ligand exchange reaction that occurs with inversion of configuration

Chart XIII

The number of sites at which ligand exchange might occur depends on the type of system involved. Systems can be envisioned in which exchange occurs at numerous sites. We will treat only the two simplest cycle types: those whose chiromers possess only one site that undergoes ligand exchange; those whose chiromers have two exchanging sites.

3.1 Ligand Exchange at One Site

Cycles that involve ligand exchange at one site are podal, and resemble those triligostats of the last section in which all reactions occur with retention of configuration. No ligand metathesis is possible in such cycles. Chart XIV outlines a hypothetical seven-reaction cycle of this type, only the first two reactions of which were completed [13]. The sequence is modeled after the actual, podal, triligostatic seven-reaction cycle also outlined in Chart XIV [14].

3.2 Ligand Exchange at Two Sites

When two sites are involved in exchange in reaction cycles with atropisomers as chiromers, the cycles can be either podal or antipodal. As in the last section, we assume that *each ligand exchange reaction occurs only with retention* of configuration. Cycles in which two different ligands interchange their sites of attachment (ligand metathesis or *LM*) possess the possibility of the cycle including two enantiomers (antipodal). Any non-reducible cycle contains either zero or one *LM*. If an *LM* is absent from the cycle, the cycle must be podal. If an *LM* is present, the cycle may be either podal or antipodal, depending on the symmetry properties of the parent system.

The parent system is that in which the two sites of exchange carry the same ligand. If the parent system is chiral, only podal cycles are possible, even when an *LM* is present. If the parent system is achiral, inclusion of an *LM* produces an antipodal cycle. Thus parent systems that belong to space groups C_2 or D_2 identify cycles that must be podal. Parent systems that belong to space groups such as C_s, S_2, C_{2h} etc. provide cycles that can be either podal or antipodal depending on the absence or presence of an *LM*. Chart XV provides generalized examples of cycles that contain *LM*'s. The cycles are identified as podal or antipodal, and the systems to which the chiromers belong are identified by the symmetry properties and space group to which the parent system belongs.

| XVIII | XIX |

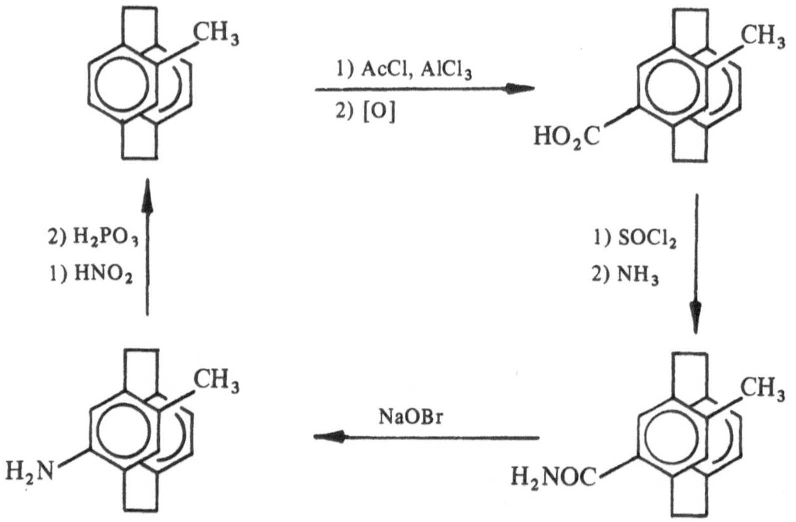
CH3
1) AcCl, AlCl3
2) [O]
CH3
HO2C
1) SOCl2
2) NH3
CH3
H2NOC
NaOBr
CH3
H2N
2) H2PO3
1) HNO2

Podal cycle with only one substitutable site

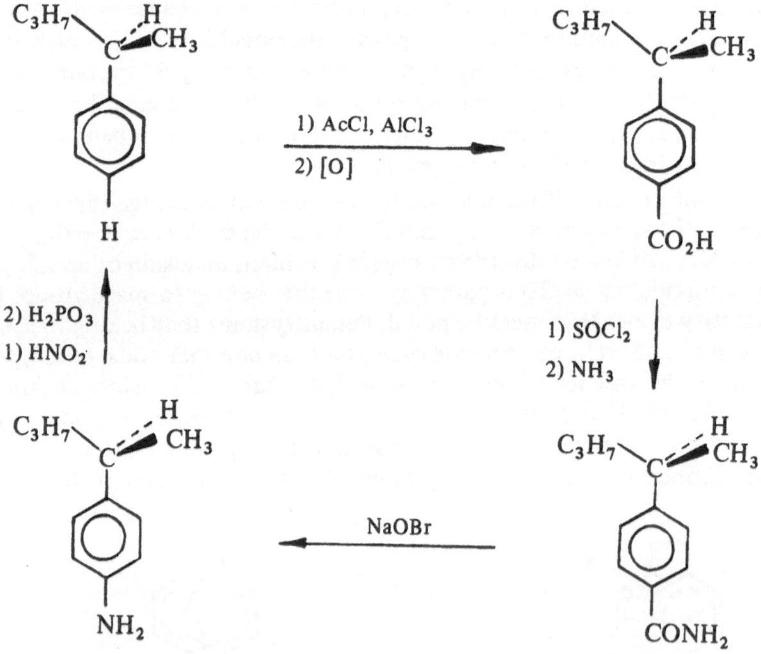
C3H7
H
CH3
C
H
1) AcCl, AlCl3
2) [O]
C3H7
H
CH3
C
CO2H
1) SOCl2
2) NH3
C3H7
H
CH3
C
CONH2
NaOBr
C3H7
H
CH3
C
NH2
2) H2PO3
1) HNO2

Podal triligostatic cycle

Chart XIV

Stereochemical cycle		Parent system		
Structure	Symmetry	Structure	Symmetry	Space group

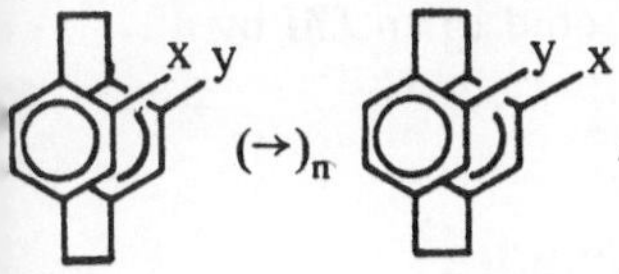	Antipodal	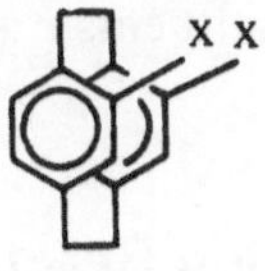	Achiral	C_s
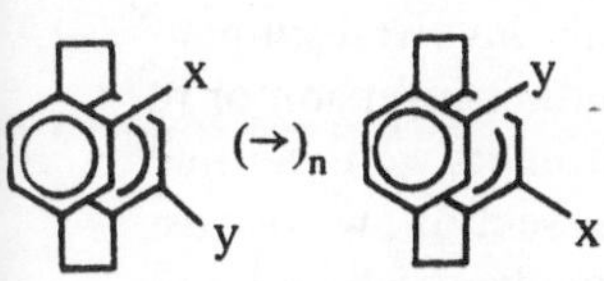	Podal	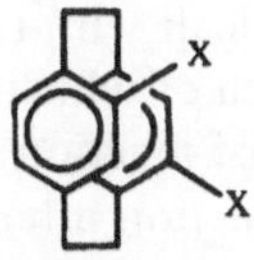	Chiral	C_2
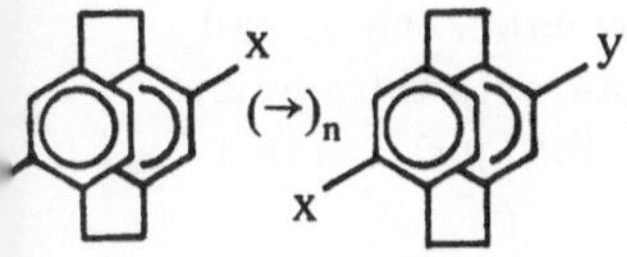	Antipodal	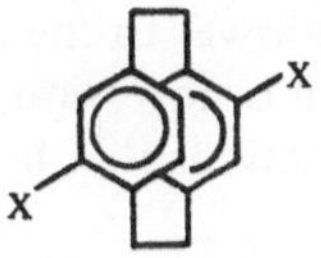	Achiral	S_2
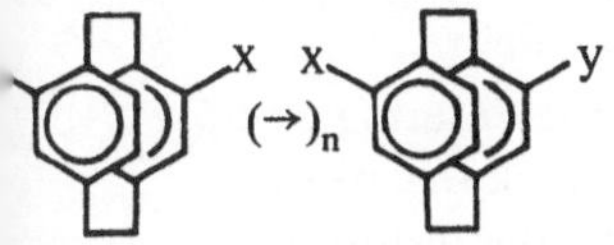	Podal		Chiral	C_2
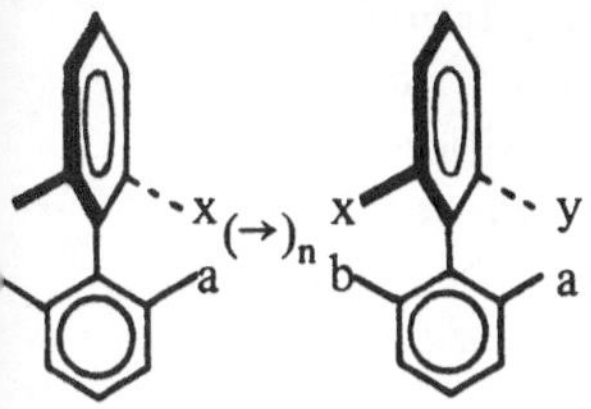	Antipodal	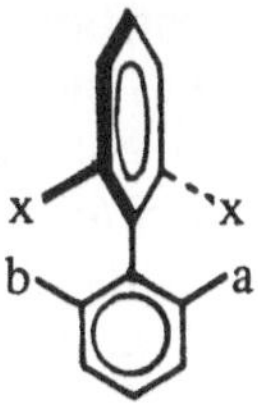	Achiral	C_s
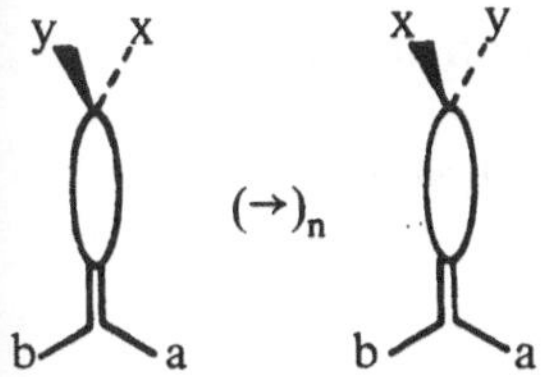	Antipodal		Achiral	C_s

Chart XV

25

Stereochemical cycles could be used to differentiate structures. For example, if an optically active molecule was known to have either structure XVIII or XIX, a choice could be made on the basis of whether a podal or antipodal cycle was generated when substituents x and y were subjected to an LM by a series of reactions.

4. Associative Substitution Reactions on Tetrahedra that Involve Rearrangeable Trigonal Bipyramids as Intermediates

In the preceding sections we dealt with reaction cycles that involved simple ligand exchange reactions, each of which occurred with either inversion or retention. To avoid difficulties of reaction mechanism, chiromers were defined as isolable compounds, not reaction intermediates. In this section, we discuss associative substitution reactions [15] on tetrahedra that involve trigonal bipyramids as intermediates. The possibility exists that these intermediates isomerize before decomposing back to tetrahedra. Such isomerizations introduce interesting complications into the answer to the question of whether the overall conversion went with retention or inversion of configuration. In this section we develop maps and generalizations that help to answer this question in a variety of cases.

4.1 Associative Substitution without Isomerization

Attack on a tetrahedron (1234) by a potential ligand (5, the entering group) in principle can occur either on a face or an edge. When attack occurs on a face, ligand 5 becomes attached to the axial position (a) of the resulting trigonal bipyramid (axial attack). When attack occurs on an edge, ligand 5 becomes attached to the equatorial position (e) of the resulting trigonal-bipyramid (equatorial attack). Ligand 4 as the leaving group of the trigonal-bipyramid in principle can leave from an axial (a) or equatorial position (e).

Mechanistic symbols such as ae refer to the positions of the leaving and entering groups on the trigonal-bipyramid. The first letter of the pair places the leaving (4), and the second the entering group's (5) position on the trigonal-bipyramid.

The *aa* process corresponds to the well-known SN_2 reaction, and proceeds with inversion. The *ee* reaction also proceeds with inversion[a]. The *ea* and *ae* reactions both proceed with retention. Chart XVI illustrates these possibilities.

---

[a] To the best of our knowledge, Westheimer and coworkers (Ref. [16,17]) were the first to recognize the *ee* route for inversion in substitution reactions on tetrahedral species.

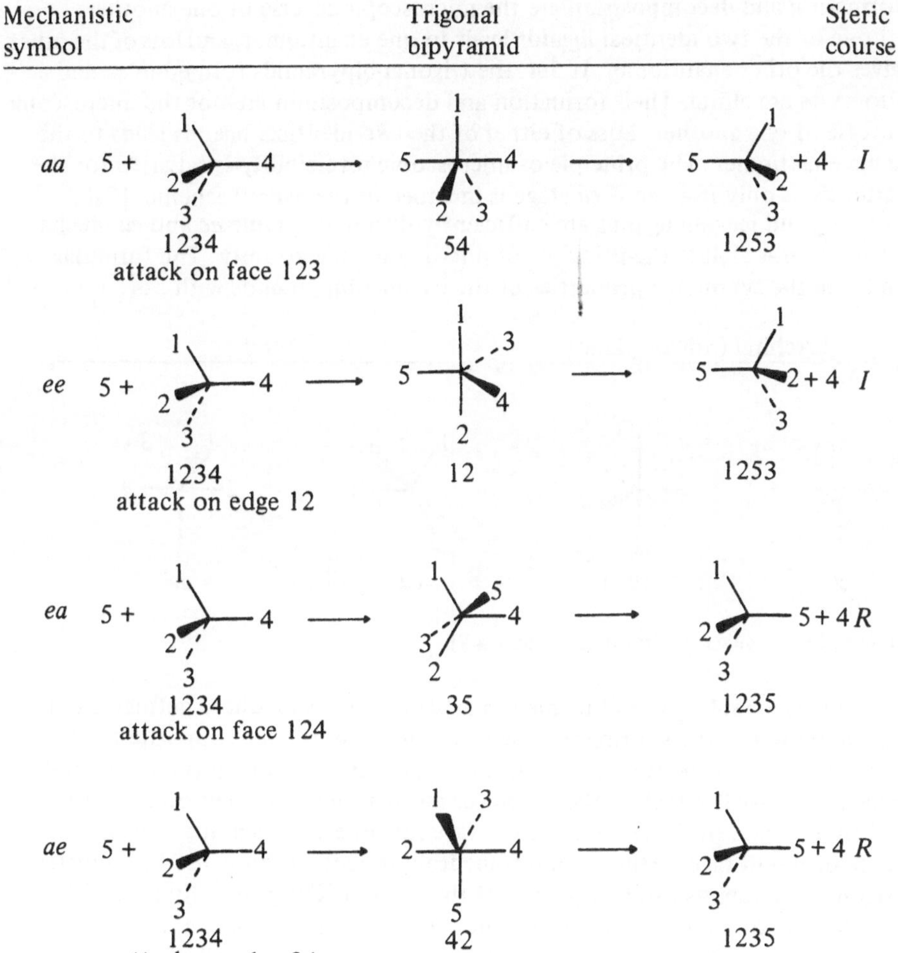

Mechanistic symbol		Trigonal bipyramid		Steric course

aa   5 + [structure] → [structure] → [structure] + 4   *I*

1234   54   1253
attack on face 123

ee   5 + [structure] → [structure] → [structure] + 4   *I*

1234   12   1253
attack on edge 12

ea   5 + [structure] → [structure] → [structure] 5 + 4 *R*

1234   35   1235
attack on face 124

ae   5 + [structure] → [structure] → [structure] 5 + 4 *R*

1234   42   1235
attack on edge 24

Chart XVI

In Chart XVI and the rest of this paper, the trigonal-bipyramids are identified by the two indices that designate the two axial ligands. Their order of
listing specifies the configuration of the trigonal-bipyramid. *The convention
adopted is the following.* The trigonal-bipyramid is viewed along the axis of
the axial ligands from the side that places the equatorial ligands in a clockwise
ascending numerical order. The index of the near axial ligand appears first,
followed by the index of the distant axial ligand.

If the entering and leaving ligands are identical (5⇄4), the trigonal-bipyramids formed in *aa* and *ee* processes are achiral (possess mirror planes). Their

27

formation and decomposition are the microscopic reverse of one another. Loss of one of the two identical ligands leads to one enantiomer, and loss of the other gives the other enantiomer. If 5≅4, the trigonal-bipyramids formed in *ae* and *ea* processes are chiral. Their formation and decomposition are not the microscopic reverse of one another. Loss of either of the two identical ligands leads to the same enantiomer. The principle of microscopic reversibility is satisfied for the latter cases only if an *ae* ⇄ *ea* stage is included in the overall scheme. If the entering and leaving ligands are sufficiently different, simple *ae* and *ea* mechanisms do not violate the principle of microscopic reversibility. The formulas indicate the symmetry properties of the trigonal-bipyramids with 5≅4.

Achiral (mirror plane)	Chiral

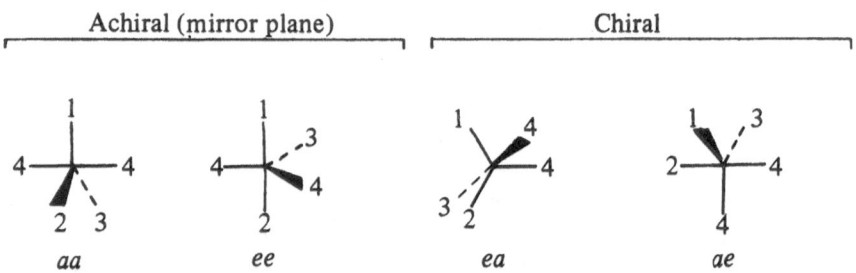

aa	*ee*	*ea*	*ae*

Trigonal-bipyramids are drawn as in Chart XVI

The trigonal-bipyramid formed in an *aa* process is uniquely defined by the configuration of the starting tetrahedron since only the face opposite the leaving group is available for axial attack. For a tetrahedron of a given configuration, edge (equatorial) attack in the *ee* and *ae* mechanisms can occur on any of three different edges. In the *ea* mechanism, face (axial) attack can occur on any of three different faces. Thus with tetrahedron of configuration 1234, ten different trigonal-bipyramids can be formed (of these Chart XVI gives only four). With a tetrahedron of configuration 1243, the ten enantiomeric trigonal-bipyramids can be formed. The twenty possible trigonal-bipyramids can all decompose directly back to their starting materials by loss of entering ligand 5, or go on to product by the loss of leaving ligand 4. Two products are possible, 1235 and 1253. Pathways

$$1234 \ (\longrightarrow)_n \ 1235$$

and

$$1243 \ (\longrightarrow)_n \ 1253$$

occur with retention, and pathways

$$1234 \ (\longrightarrow)_n \ 1253$$

and

$$1243 \ (\longrightarrow)_n \ 1235$$

occur with inversion. Clearly a map is needed to monitor the configurational changes that occur when the two enantiomeric starting tetrahedra go to the twenty trigonal bipyramids, which in turn decompose to the enantiomeric tetrahedral products. Chart XVII provides such a map.

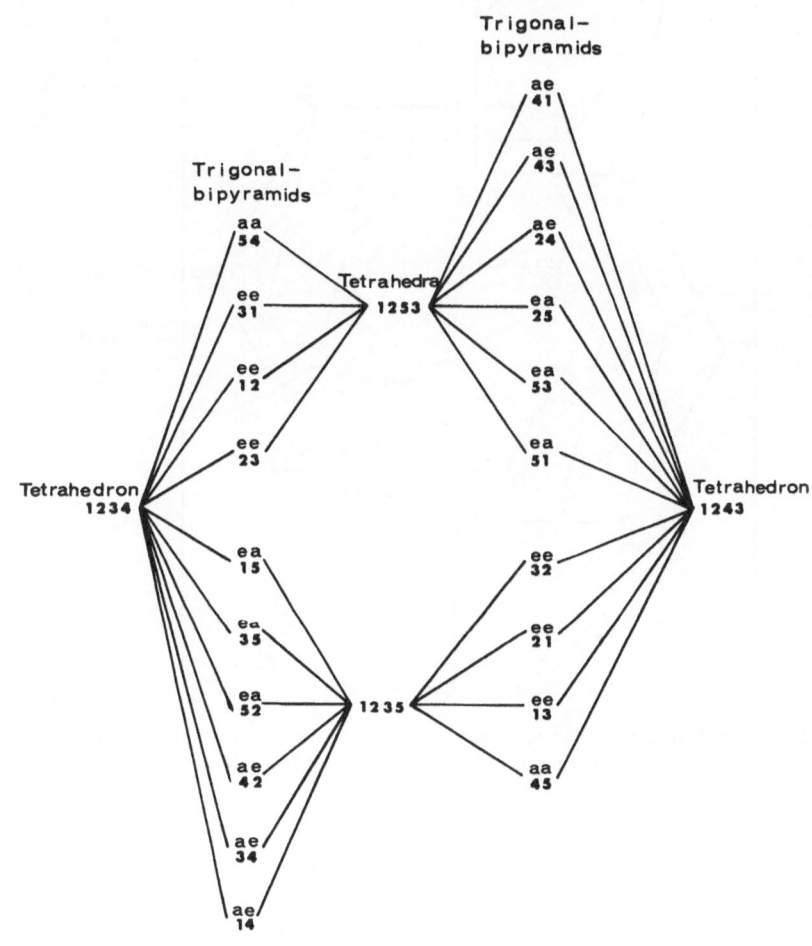

Chart XVII

4.2 Associative Substitution with Isomerization

When associative substitution of a tetrahedron is accompanied by either Berry pseudorotations [18] or turnstile isomerizations [19] of trigonal-bipyramidal in-

29

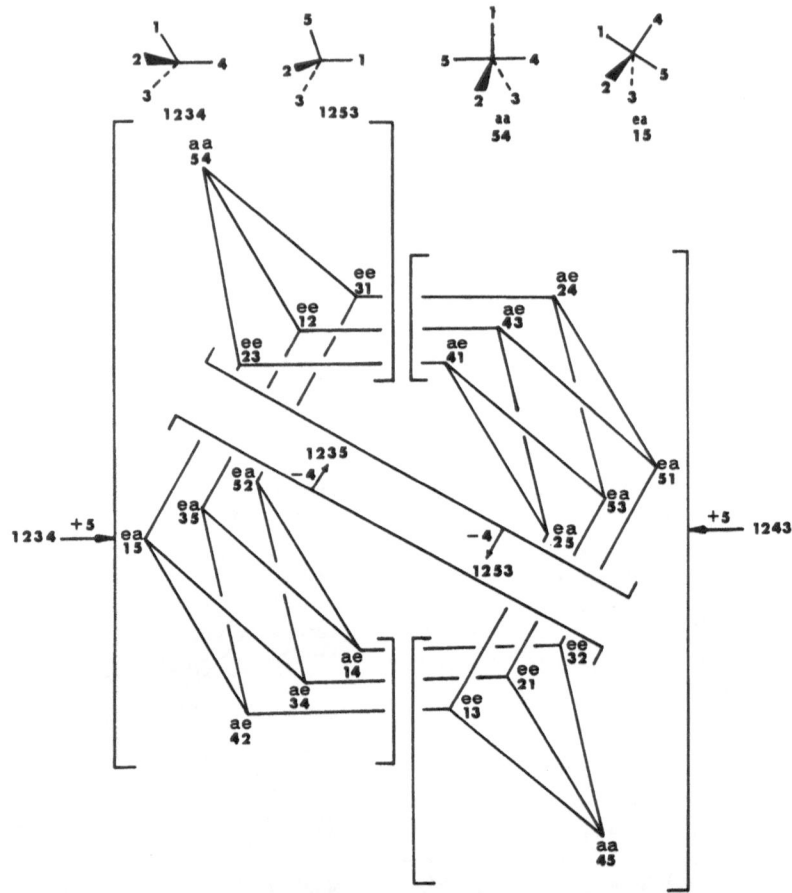

Fig. 7. Map of stereochemical courses of associative substitution of tetrahedron 1234 or 1243 by ligand 5 to form trigonal bipyramids capable of pseudorotating and decomposing to tetrahedron 1235 or 1253 by loss of ligand 4

termediates, the overall stereochemical course of the reaction is controlled by two factors. The first involves the direction of entry and departure of the entering and leaving ligands (5 and 4, respectively). The second concerns the isomerizations that the trigonal-bipyramid intermediates undergo before decomposing to tetrahedra. Fig. 7 maps all possible routes, and takes account of both factors. Although maps for pseudorotations have been published previously [4,20a-e], this is the first that combines all substitutions with all pseudorotations.

The map of Fig. 7 provides symbols for the four possible tetrahedra (four indices) and twenty possible trigonal-bipyramids (two indices). The latter are

placed at the intersections of the three lines. As in the last section, the two indices specify the axial ligands, and their order indicates the configuration of
the bipyramid. The first of the two letters above each pair of indices indicates
the position of the leaving ligand (4) to be either axial (*a*) or equatorial (*e*).
The second of the two letters specifies the position of the entering ligand (5)
as axial (*a*) or equatorial (*e*).

In the map of Fig. 7, each line connecting two intersections represents an
isomerization to a different configuration. If the isomerization is a pseudorotation [18] (see below), the pivotal ligand is the one not numbered in either the
indices of the starting or final trigonal-bipyramids. If the isomerization is an
Ugi turnstile [19] (see below), the unique ligand that remains equatorial during
the isomerization is the one not numbered in the indices of either starting
material or product.

Chart XVIII provides an example of a pseudorotation that identifies the
pivotal ligand as apical in the square-pyramidal transition state. An Ugi turnstile isomerization is also illustrated, and the unique ligand that remains equatorial during the isomerization is identified.

Pseudorotation

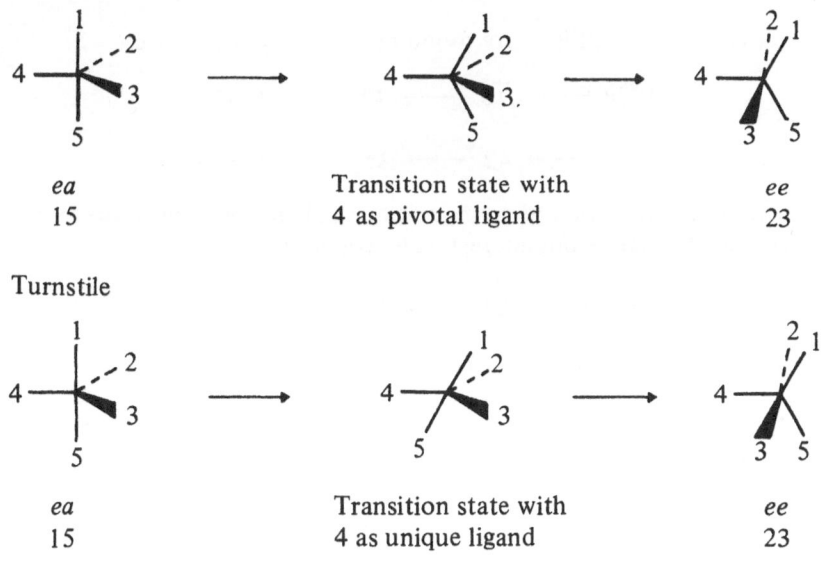

Turnstile

Chart XVIII

The map of the trigonal-bipyramid interconversions of Fig. 7 has a center
of inversion and other interesting symmetry properties. All of the bipyramids

31

in the left half of the map are formed by addition of ligand 5 to the left tetra-hedron, 1234. All of the bipyramids in the right half of the map are formed by addition of ligand 5 to the right tetrahedron, 1243. The vertical brackets em-phasize these generalizations. All of the bipyramids in the lower left half of the map upon loss of ligand 4 give the upper central tetrahedron, 1235. All of the bipyramids in the upper right half of the map decompose by loss of ligand 4 to give the lower central tetrahedron, 1253. The diagonal brackets point to these generalizations. Thus those isomerizations traced by lines *only* in the lower-left half, or only in the upper right half of the map do not alter the stereochemical course of a substitution reaction. However, inclusion of a pseu-dorotation or turnstile isomerization that connects the lower-left to the upper-right region has the effect of inverting the overall stereochemical course of a substitution reaction. All of the six diagonal lines (broken by brackets) that connect the lower-left with the upper-right half of the map represent pseudo-rotations about the leaving group (4) as the pivotal ligand. Similarly, these lines represent turnstile isomerizations with the leaving group (4) as the unique ligand (the one that remains equatorial). The following generalization emerges:

In substitution reactions on chiral tetrahedra that involve trigonal-bipyr-amids as intermediates, only isomerizations with the leaving group as a pivotal (or unique) ligand affect the stereochemical course of the overall substitution reaction.

Examples of the use of Fig. 7 are found in Chart XIX. Sequences

$$1234 \longrightarrow 15 \longrightarrow 42 \longrightarrow 1235$$

and

$$1234 \longrightarrow 42 \longrightarrow 15 \longrightarrow 1235$$

proceed with overall retention of configuration, and are confined to the lower left quadrant of the map. More interesting is sequence

$$1234 \longrightarrow 23 \longrightarrow 15 \longrightarrow 1235$$

which proceeds with retention, in effect by two inversions merging in the se-quence. One of these inversions is associated with the *ee* substitution, and the other with the isomerization around the leaving group (23 $\xrightarrow{4}$ 15). Both of the left quadrants of the map are involved in the overall sequence. Sequen-ces

$$1234 \longrightarrow 54 \longrightarrow 23 \longrightarrow 1253$$

and

$$1234 \longrightarrow 23 \longrightarrow 54 \longrightarrow 1253$$

both go with inversion because they involve *aa* or *ee* substitution processes (see Chart XVI). Both reaction paths are confined to the upper left quadrant of the map. Sequence

$$1234 \longrightarrow 15 \longrightarrow 23 \longrightarrow 1253$$

also occurs with inversion, which is associated with the isomerization around the leaving group ($15 \xrightarrow{4} 23$). Both of the left quadrants of the map are involved in the overall sequence.

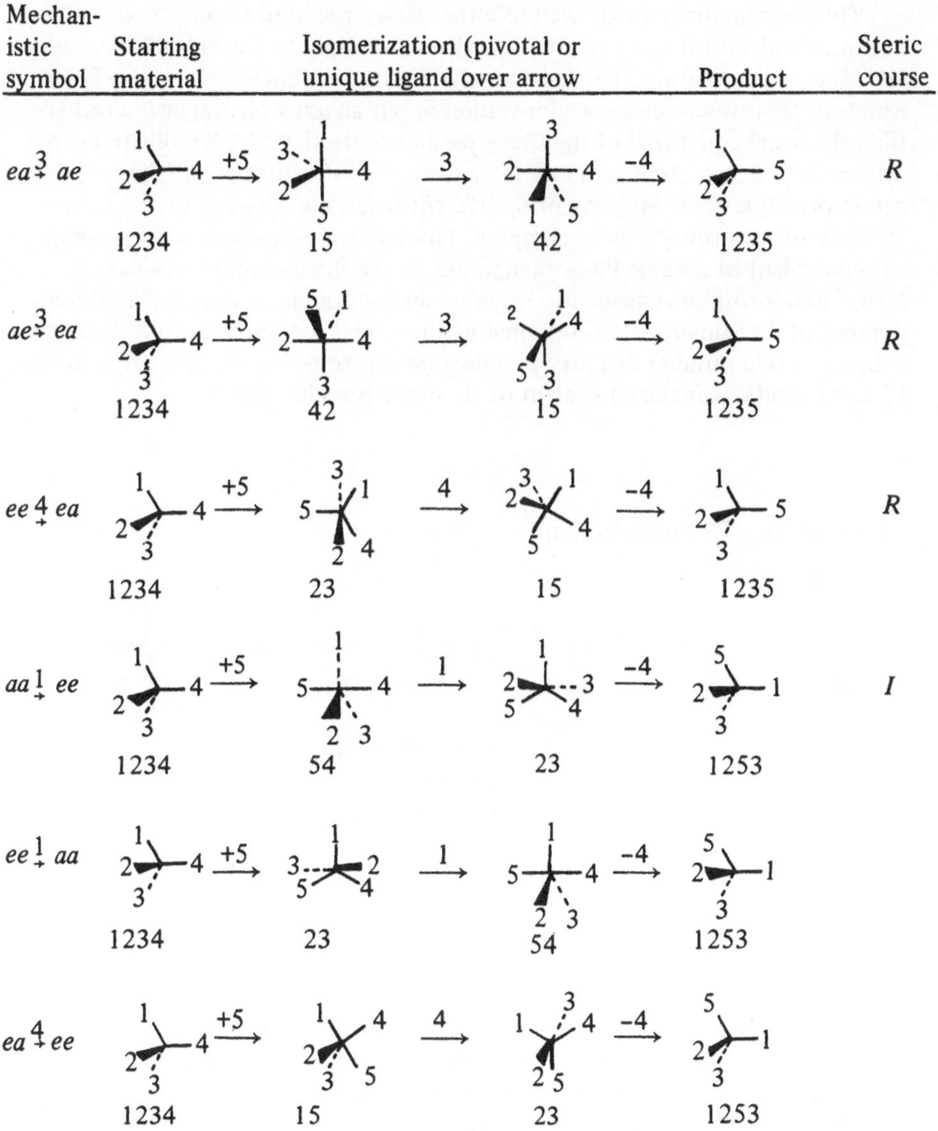

Mechanistic symbol	Starting material	Isomerization (pivotal or unique ligand over arrow			Product	Steric course

Chart XIX

33

The trigonal-bipyramid part of this map is useful for describing the pseudo-rotations or turnstile isomerizations of pentacoordinate chiral compounds. A minimum of five reactions are needed to convert a bipyramid to its enantiomer. One of these five reactions must involve 4 as the pivotal or unique ligand. Only an odd number of reactions can be used to get from a bipyramid to its enantiomer, and one of them must involve 4 as the pivotal or unique ligand.

This map applies equally well to either Berry pseudorotations or Ugi turnstile mechanisms for isomerizations. It does not apply to Gielen's P3 mechanism [21] (Muetterties Process 2)[22] for isomerization of trigonal-bipyramids. These isomerizations resemble a pseudorotation about an *axial* pivotal ligand rather than the usual equatorial of the Berry pseudorotation. Chart XX illustrates the process. The P3 reactions number 60, and each of the 20 trigonal-bipyramids can in principle enter into any of 6 different reactions. Maps of the P3 isomerizations are accordingly more complex. Three Berry pseudorotations accomplish the equivalent of a single P3 isomerization. Since the minimum number of Berry pseudorotations needed to generate an enantiomer is five, the minimum number of P3 isomerizations required is three. This relationship holds because it takes an odd number of Berry pseudorotations to get to enantiomer, and two P3 isomerizations is the equivalent of six Berry pseudorotations.

Example of a P3 isomerization

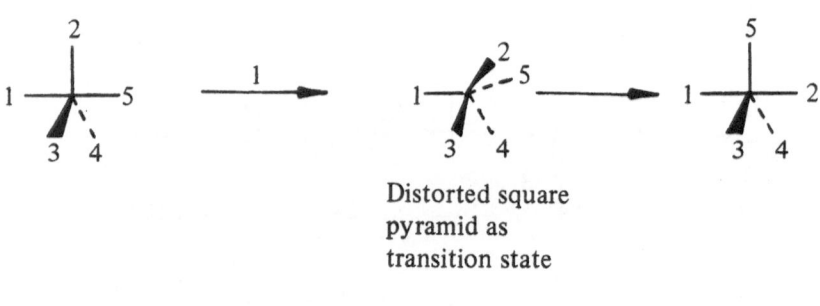

Distorted square
pyramid as
transition state

Chart XX

A map that describes associative substitution of chiral tetrahedra to give isomerizable square-pyramids has been described [4]. A diamond lattice map with an internal network has also been developed [4] that traces the Berry pseudorotations of chiral trigonal-bipyramids. This earlier paper provided the same generalizations as did the more detailed map of Fig. 7.

4.3 Restrictions on Direct Substitution or Formation of Trigonal Bipyramid Intermediates

Structural limitations as to modes of attack to form a trigonal bipyramid either as an intermediate or a transition state depend on the character of the four ligands attached to the tetrahedral center, and to a lesser degree on the properties of the attacking group. Whether the attack by a nucleophile occurs on a face (axial) or on an edge (equatorial) is determined by the electronegativity, bulk or bond angle requirements of the five substituents of the trigonal bipyramid formed. According to the specific properties involved, attack may be limited to certain of the six edges or four faces. To express it in another way, it is possible that in the trigonal bipyramid formed one or more groups may be restricted to an axial or an equatorial position.

The restrictions fall into two categories:

1) A given position is required for one or more groups when the magnitude of one of their properties differs greatly from that of the other attached groups. Thus the trigonal bipyramid formed must have the restricted groups in their most compatible positions, axial or equatorial. This situation arises when large differences in electronegativities or bulk exist among the substituents.

2) Two groups are restricted as to their relative positions, but the individual positions may be interchangeable. This limitation arises when the bond angle between groups is sufficiently limited by incorporation of those groups in a ring system. Obviously the two categories may exist simultaneously.

An exhaustive discussion of restrictive situations is beyond the scope of our treatment. Position preference has been examined in pseudorotations of stable five-coordinate phosphorus compounds [17b,23,24]. The generalities derived from these studies may possibly be applied to formation of trigonal bipyramids as well.

1) Strongly electronegative groups occupy available axial positions; strongly electropositive groups the equatorial positions.

2) Rings smaller than 9–12 members cannot span the diaxial positions. Four-membered rings must occupy axial-equatorial positions. Five-membered rings strongly prefer to be axial-equatorial. Six-membered rings prefer to be diequatorial. Bicycloheptyl and bicyclooctyl systems need to span two equatorial and one axial positions. Attack on an edge between two substituents common to the same ring, making them diaxial in the trigonal bipyramid formed, is prohibited. Attack on a face containing two ring substituents, making them diequatorial, is unlikely with five-membered and inhibited with four-membered rings.

3) Axial bonds being richer in p character are longer than equatorial bonds. Axial ligands have three near neighbors whereas equatorial ligands have four, two somewhat less near. It is likely that the space available for bulky, axial attack may be the only route. Attack on any of the three faces having the

bulky group as a common apex is a high energy route since the trigonal bipyramid would contain the bulky group in an equatorial position. Whether the trigonal bipyramid is a transition state or an intermediate, the preferred route for electronegative attacking and leaving groups seems to be *aa* to give inversion of configuration. One-stage substitutions on carbon are generally accepted to go *aa* through a transition state in which carbon is approximately hybridized sp^2-p. Nucleophilic substitution at second row elements involves intermediates or transition states in which the central atom is hybridized sp^3d. The subsequent discussion is limited to second row elements.

An abundance of examples of nucleophilic substitution reactions are available which occur with inversion. Examples for phosphorus and silicon are found in Charts XI and XII respectively. A fine illustration of what are probably *aa* inversion mechanisms at sulfur that involve electronegative entering and leaving groups is found in Chart XXI [25a-e].

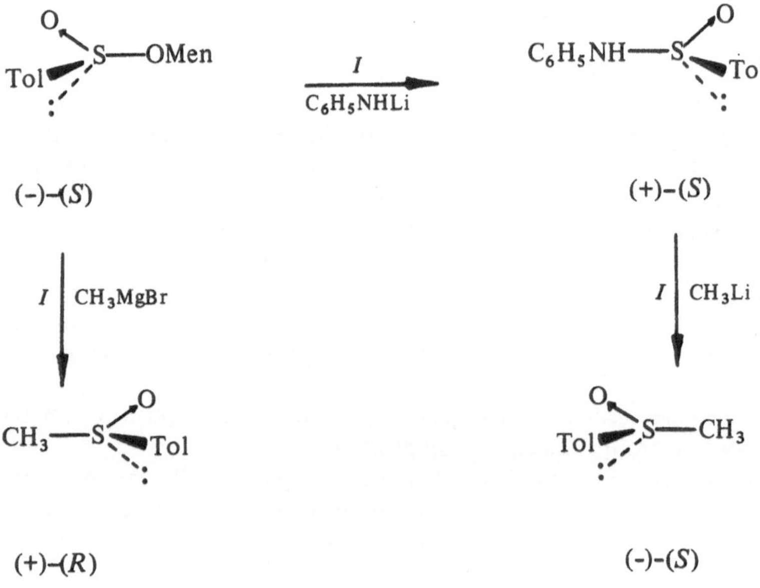

Tol = p-CH$_3$C$_6$H$_4$; Men = menthyl

Chart XXI

Two examples of inversion are known to occur when an *aa* mechanism is unavailable, the best possible route then being *ee*. These are formulated in Chart XXII. In the first case the leaving and entering groups belong to the same six-membered ring [4]. In the second [26] the presence of a four-membered ring in the starting material enforces an *ee* mechanism in spite of the fact that the unshared electron pair is forced into an axial position in the trigonal bipyramid. An analogous reaction at phosphorus has been reported in which an *ee*

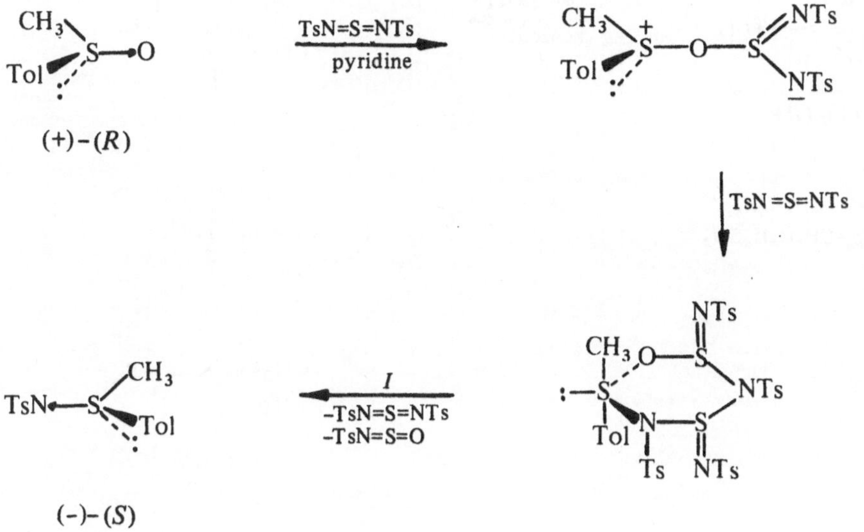

$Tol = p\text{-}CH_3C_6H_4;$ $Ts = p\text{-}CH_3C_6H_4SO_2$

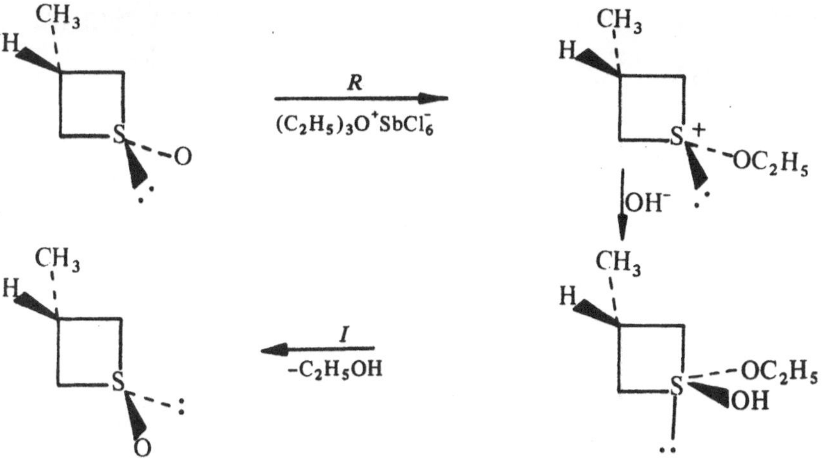

Chart XXII

mechanism is enforced by the presence of a four-membered ring [27]. Haake and Westheimer were the first to postulate an *ee* substitution in the hydrolysis of cyclic phosphate esters [16].

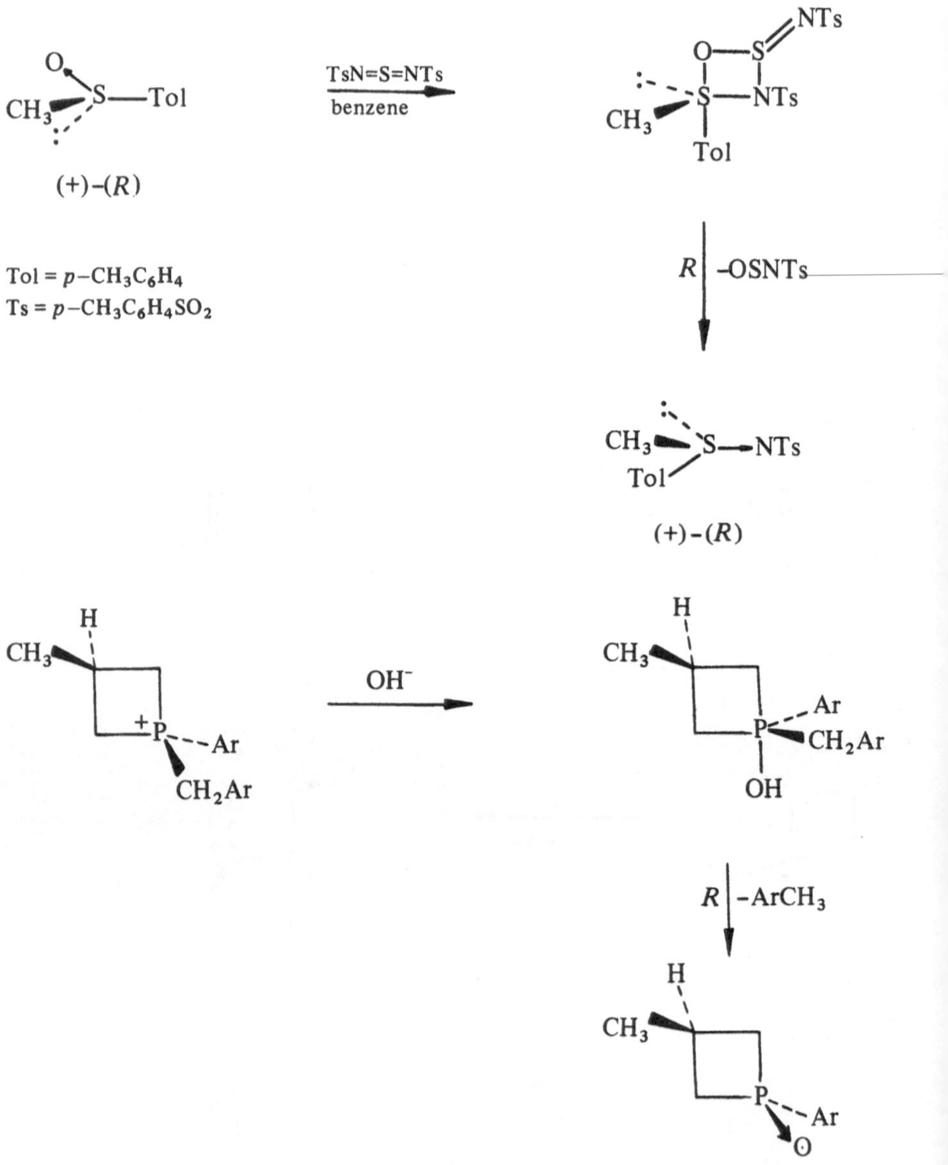

TsN=S=NTs
benzene
(+)-(R)
Tol = p-CH3C6H4
Ts = p-CH3C6H4SO2
R
-OSNTs
(+)-(R)
OH⁻
R
-ArCH3

Chart XXIII

Several examples are available for substitution at silicon [11], sulfur and phosphorus in which the reaction occurs with retention of configuration. Usually these reactions do not involve highly electronegative attacking reagents or highly electronegative leaving groups. In some cases one can postulate mechanisms in which the entering and leaving groups form a small ring as an intrinsic part of the substitution mechanism. An illustration of this is found in Chart XXIII in which a sulfoxide is converted to a sulfimide by tosylated sulfodiimide. This substitution occurs with retention in benzene [28a,b], and is found to be monomolecular in sulfodiimide [28b]. This reaction should be contrasted with that between the same reagents in Chart XXII. The latter reaction [4] is conducted in pyridine, is bimolecular in sulfodiimide, and occurs with inversion of configuration by an *ee* mechanism. The second example of Chart XXIII makes use of the *ea* retention mechanism induced because a non-leaving group as a member of a four-membered ring must occupy an axial position [29].

McEwen and Van der Werf long ago demonstrated [30] that arylbenzyldialkylphosphonium salts when treated with strong base produced the phosphine oxide with inversion of configuration in a reaction in which benzyl served as leaving group. In contrast Trippett, *et al.* found that the presence of a *t*-butyl group [31] in place of a smaller alkyl group directed the reaction to a retention stereochemical course (see Chart XXIV). This changeover in mechanism may reflect the preference of the bulky *t*-butyl group for the less crowded axial position in the trigonal-bipyramid. When two *t*-butyl groups were present in the phosphonium salt, the reaction to form phosphine oxide with base became very very slow. The benzyl no longer served as leaving group, but a *t*-butyl group was expelled. Although the trigonal-bipyramid is formulated here with the two *t*-butyl groups axial, it is also possible that an axial-equatorial arrangement of these large groups provides a better route.

The nature of pseudorotation restrictions of stable five-coordinate phosphorus compounds has been studied [17b,23,24]. Pseudorotations which move groups into restricted positions do not occur. Pseudorotations involving axial-equatorial rings occur by interchange of the axial for the equatorial group. However, only those pseudorotations about the leaving group as pivot alter the stereochemical outcome of substitution.

Two cases are known of conversion of

tetrahedron → trigonal-bipyramid → isomerized-trigonal-bipyramid →
→ tetrahedron

in which the pseudorotation step appears to be necessary in the mechanism. Racemization of (+)-methylphenylpropylphosphine oxide [32] by lithium aluminum hydride occurs before reduction. The mechanism postulates equatorial attack of hydride, pseudorotation about hydride as pivot, and subsequent departure of hydride equatorially to form the enantiomer of the original tetrahedron (Chart XXV).

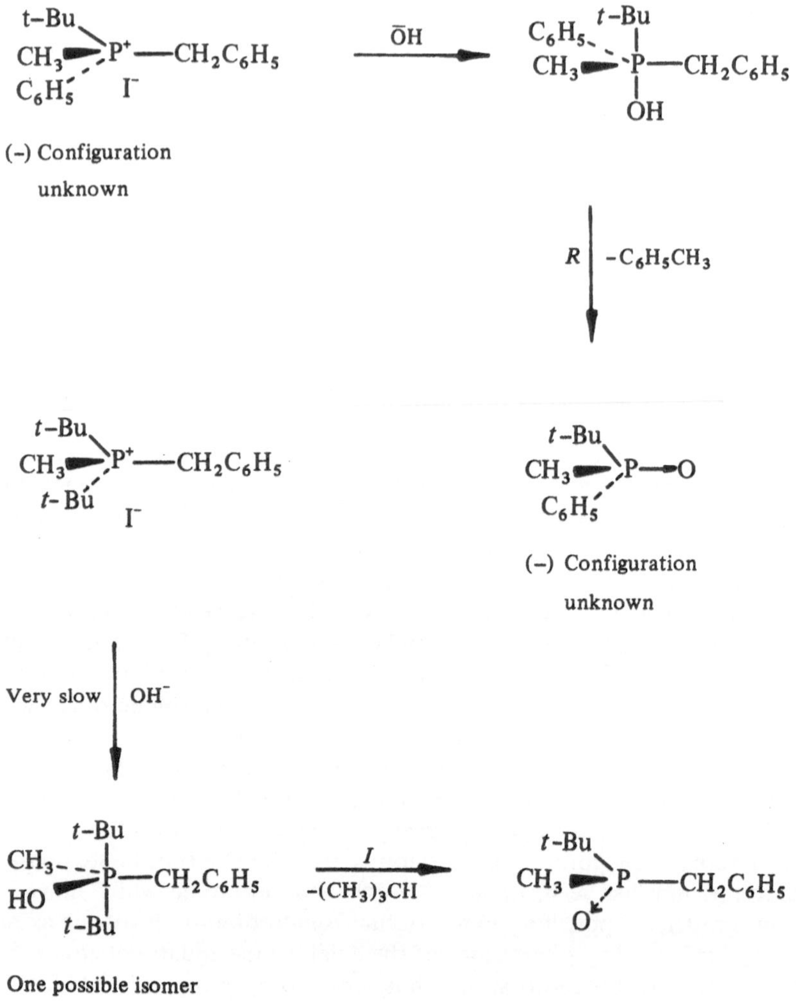
t-Bu
CH3
P+
CH2C6H5
C6H5
I-
(-) Configuration unknown
OH
C6H5
t-Bu
CH3
P
CH2C6H5
OH
R
-C6H5CH3
t-Bu
CH3
P+
CH2C6H5
t-Bu
I-
t-Bu
CH3
P
O
C6H5
(-) Configuration unknown
Very slow
OH-
t-Bu
CH3
P
CH2C6H5
HO
t-Bu
One possible isomer
I
-(CH3)3CH
t-Bu
CH3
P
CH2C6H5
O

Chart XXIV

Pseudorotation, *ea* → *ae*, is invoked to satisfy microscopic reversibility in the retention result [20e)] observed in the ethoxide displacement by hydroxide on a phosphorus center when a four-membered ring prevents the *aa* inversion mechanism (Chart XXV). The same mechanism was originally proposed by Westheimer and Dennis for the acid-catalyzed hydrolysis of cyclic phosphate esters [17a)].

Our ambition in this paper has been to provide a framework in terms of which diverse stereochemical reaction cycles can be followed. The maps devel-

One possible edge attack

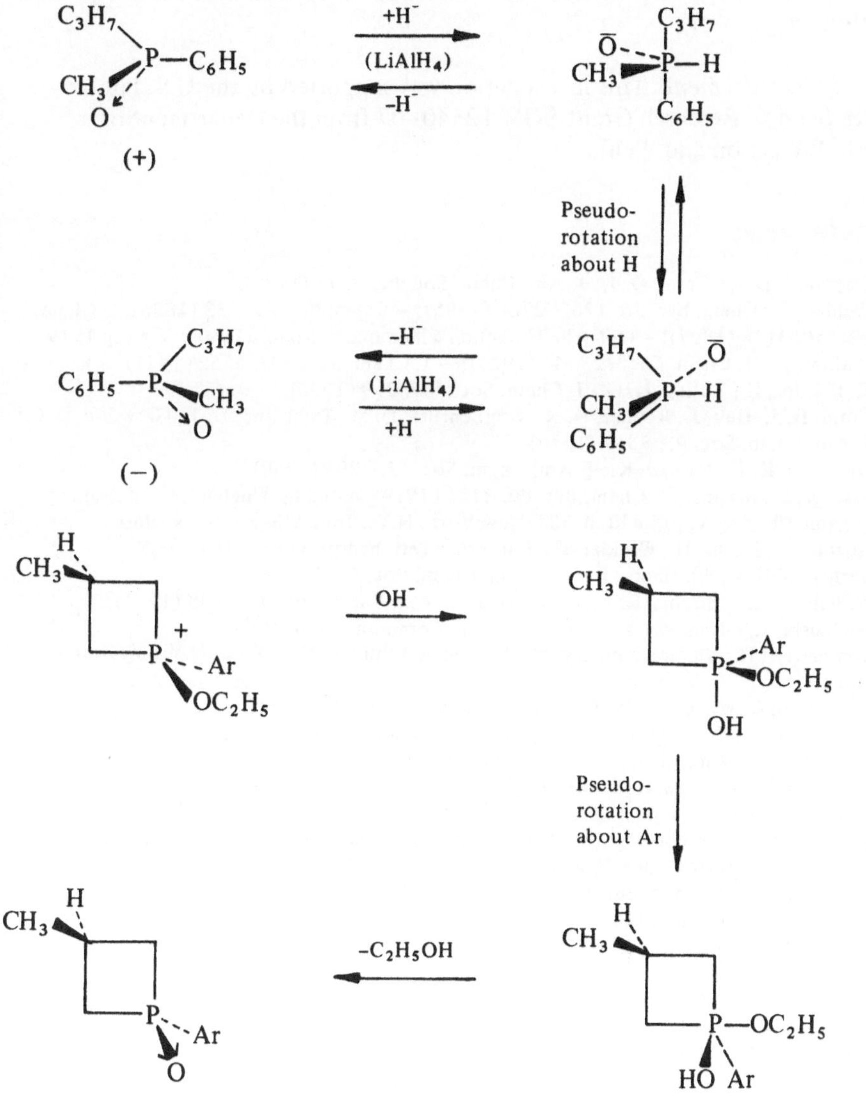

$$C_3H_7 \diagdown \atop CH_3 \diagup P-C_6H_5, \quad O$$
(+)

$$\xrightarrow[\text{--H}^-]{\substack{+H^- \\ (LiAlH_4)}}$$

$$\overline{O}\cdots P-H \atop C_6H_5, \quad CH_3, \quad C_3H_7$$

Pseudo-
rotation
about H

$$C_6H_5-P \diagup{}^{C_3H_7} \diagdown{}_{CH_3}, \quad O$$
(−)

$$\xleftarrow[\substack{+H \\ }]{\substack{-H^- \\ (LiAlH_4)}}$$

$$C_3H_7 \diagdown \overline{O} \atop CH_3 \diagup P-H \atop C_6H_5$$

$$CH_3 \cdots \boxed{} \atop H \\ P \diagdown{-Ar} \atop{}^{+}{OC_2H_5}$$

$$\xrightarrow{OH^-}$$

$$CH_3 \cdots \boxed{} \atop H \\ P \diagdown{Ar} \atop{}{OC_2H_5} \\ OH$$

Pseudo-
rotation
about Ar

$$CH_3 \cdots \boxed{} \atop H \\ P \diagdown{Ar} \atop O$$

$$\xleftarrow{-C_2H_5OH}$$

$$CH_3 \cdots \boxed{} \atop H \\ P-OC_2H_5 \atop HO\ Ar$$

Chart XXV

oped in the first three sections can be modified to accommodate the restrictions discussed in the fourth section. Simple omission from the maps of appropriate lines connecting chiromers provides restricted maps which can account for the remaining possible reaction routes. An example of how this may be done is found in the work of Gorenstein and Westheimer as applied to restricted pseudorotation [24].

Acknowledgement: This investigation was supported by the U.S. Public Health Service, Research Grant #GM 12540-07 from the Department of Health, Education and Welfare.

5. References

[1] Garwood, D. C., Cram, D. J.: J. Am. Chem. Soc. *92*, 4575 (1970).
[2] Walden, P.: Chem. Ber. *28*, 1287, 2766 (1895); – Chem. Ber. *29*, 133 (1896); – Chem. Ber. *30*, 3146 (1897); – Optische Umkehrerscheinungen. Braunschweig: Vieweg 1919.
[3] Phillips, H.: J. Chem. Soc. *123*, 44 (1923); – J. Chem. Soc. *127*, 2552 (1925); – Kenyon, J., Phillips, H., Turley, H. G.: J. Chem. Soc. *127*, 399 (1925).
[4] Cram, D. J., Day, J., Rayner, D. R., von Schriltz, D. M., Duchamp, D. J., Garwood, D. C.: J. Am. Chem. Soc. *92*, 7369 (1970).
[5] De Bruin, K. E., Mislow, K.: J. Am. Chem. Soc. *91*, 7393 (1969).
[6] Fischer, E., Brauns, F.: Chem. Ber. *47*, 3181 (1914), noted by Wheland, H.: Advanced Organic Chemistry, 3rd edit. p. 305. New York, N.Y.: John Wiley & Sons 1960.
[7] Horner, L., Fuchs, H., Winkler, H., Rapp, A.: Tetrahedron Letters *1963*, 965.
[8] Bernstein, H. I., Whitmore, F. C.: J. Am. Chem. Soc. *61*, 1324 (1939).
[9] Williams, T. R., Booms, R. E., Cram, D. J.: J. Am. Chem. Soc. *93*, 7338 (1971).
[10] Michalski, J., Okruszek, A., Stec, W.: Chem. Commun. *1970*, 495.
[11] Sommer, L. H.: Stereochemistry, Mechanism and Silicon. New York, N.Y.: McGraw-Hill 1965.
[12] – Michael, K. W., Korte, W. D.: J. Am. Chem. Soc. *89*, 868 (1967).
[13] Delton, M. H., Cram, D. J.: J. Am. Chem. Soc. *92*, 7623 (1970).
[14] Cram, D. J.: J. Am. Chem. Soc. *74*, 2152 (1952).
[15] Langford, C. H., Gray, H. B.: Ligand Substitution Processes, p. 8. New York, N.Y.: W. A. Benjamin 1965.
[16] Haake, P. C., Westheimer, F. H.: J. Am. Chem. Soc. *83*, 1102 (1961).
[17a] Dennis, E. A., Westheimer, F. H.: J. Am. Chem. Soc. *88*, 3432 (1966).
[17b] Westheimer, F. H.: Accounts Chem. Res. *1*, 70 (1968).
[18] Berry, R. S.: J. Chem. Phys. *32*, 933 (1960).
[19] Ugi, I., Marquarding, D., Klusacek, H., Gokel, G., Gillespie, P.: Angew. Chem. Intern. Ed. *9*, 703 (1970).
[20a] Balaban, A. T., Fǎrcasin, D., Bǎnicǎ, R.: Rev. Roum. Chim. *11*, 1205 (1966).
[20b] Dunitz, J. D., Prelog, V.: Angew. Chem. Intern. Ed. *7*, 725 (1968).
[20c] Lauterbur, P. C., Ramirez, F.: J. Am. Chem. Soc. *90*, 6722 (1968).
[20d] Muetterties, E. L.: J. Am. Chem. Soc. *91*, 1636, 4115 (1969).
[20e] De Bruin, K. E., Naumann, K., Zon, G., Mislow, K.: J. Am. Chem. Soc. *91*, 7031 (1969).
[21] Gielen, M., Vanlautem, N.: Bull. Soc. Chim. Belges *79*, 679 (1970).
[22] Muetterties, E. L.: J. Am. Chem. Soc. *91*, 4115 (1969).
[23] Ramirez, F.: Accounts Chem. Res. *70*, 168 (1968).
[24] Gorenstein, D., Westheimer, F. H.: J. Am. Chem. Soc. *92*, 634 (1970).
[25a] Andersen, K. K.: Tetrahedron Letters *1962*, 93.
[25b] Mislow, K., Green, M. M., Laur, P., Melillo, J. T., Simmons, T., Ternay, A. L., Jr.: J. Am. Chem. Soc. *87*, 1958 (1965).

[25c)] Jacobus, J., Mislow, K.: Chem. Commun. *1968*, 253.
[25d)] Colonna, S., Giovini, R., Montanari, S.: Chem. Commun. *1968*, 865.
[25e)] Nudelman, A., Cram, D. J.: J. Am. Chem. Soc. *90*, 3869 (1968).
[26)] Tang, R., Mislow, K.: J. Am. Chem. Soc. *91*, 5644 (1969).
[27)] Smith, D. J. H., Trippett, S.: Chem. Commun. *1969*, 855.
[28a)] Christensen, B. W.: Chem. Commun. *1971*, 597.
[28b)] Yamagishi, F., Cram, D. J.: unpublished results.
[29)] Corfield, J. R., Shutt, J. R., Trippett, S.: Chem. Commun. *1969*, 789.
[30)] Kumli, K. F., McEwen, W. E., Van der Werf, C. A.: J. Am. Chem. Soc. *81*, 3805 (1959); McEwen, W. E., Kumli, K. F., Bladé-Font, A., Zanger, M., Van der Werf, C. A.: J. Am. Chem. Soc. *86*, 2378 (1964).
[31)] De'Ath, N. J., Trippett, S.: Chem. Commun. *1969*, 172.
[32)] Henson, P. D., Naumann, K., Mislow, K.: J. Am. Chem. Soc. *91*, 5645 (1969).

Received September 3, 1971

Quantum Biochemistry
at the All- or Quasi-All-Electrons Level

Dr. Alberte Pullman

Institut de Biologie Physico-Chimique, Laboratoire de Biochimie Théorique associé au C.N.R.S. Paris, France

Contents

A. Pullman

I. Introduction

The last decade has seen an unprecedented development in the possibilities of applying the methods of quantum mechanics to the study of molecular structure. Increased computing facilities have speeded up the elaboration of new, rapid, semi-empirical techniques for dealing with all the valence-shell electrons of large molecules, while almost simultaneously there has been a great advance in the applicability of non-empirical methods for *all* the electrons.

The impact of these improvements in the quantum mechanical tools on the development of quantum biochemistry has been considerable, the reason for this lying in the nature of the molecules involved in biochemical problems: most of the building blocks of living matter are large organic compounds, often linked together into huge polymers or into smaller though quite complex "associations" of different kinds.

The fortunate fact that a number of the biochemical units are conjugated heterocycles has permitted the first theoretical unraveling of problems connected with their electronic structure in the framework of the π-electron approximation. However simple the procedures utilized, a careful analysis of the results has allowed the interpretation of a considerable body of experimental facts as well as a number of predictions later confirmed by subsequent experimentation [1]. Although these results have survived the test of the successive refinements of the π-electron theories and have been complemented by the introduction of a simple representation of the σ-framework, the possibilities of treating the σ and π electrons *simultaneously* on an equal footing had to be explored in order to establish the theory on a firmer basis and also to gain further insight into some fine features of electronic properties which are otherwise inaccessible. Thus, the first outcome of the penetration of all-valence and all-electrons methods into biochemistry has been to deepen and refine previous studies.

But another benefit of the all-electrons era in the biochemical domain, and perhaps the most essential one, lies in the fact that it has opened up new fields of investigation: for instance, the entirely intact domain of saturated, or partially saturated biomolecules and the whole area of *conformational problems* which is of central importance, not only for the understanding of the behavior of large polymers, but also for questions related to coenzyme structure, steroids, vitamins and drug action, etc. Last but not least, the field of *molecular associations* governed by hydrogen-bonding, charge-transfer, Van der Waals-London "forces", etc. which has until now been the preserve of semi-empirical perturbative approaches, becomes amenable to non-truncated calculations.

It is clear that, in a review like this, some choice has to be made. We have chosen to illustrate the main lines of recent development by means of some outstanding examples, indicating in tables and lists of references what cannot be described in detail.

46

II. All-Valence and All-Electrons Calculations on Fundamental Biomolecules

Computations on biomolecules have been performed by practically all the available all-valence electrons prcedures. The simplest is the extended Hückel theory (EHT) in the version developed by Hoffmann [2] which is an extension of the well-known Hückel approximation for π electrons in the sense that the molecular orbitals are obtained as the eigenvalues of an effective Hamiltonian H that is not explicit, the matrix elements of H being treated as empirical input characteristics of the atoms involved.

Various iterative refinements of the procedure are sometimes preferred.

The second most popular all-valence electrons procedure is the CNDO/2 method [3], generalizing the zero-differential overlap Pariser-Parr hypothesis to an otherwise rationalized SCF scheme. A parallel between EHT, iterative EHT and CNDO/2 has been drawn elsewhere [4,5].

Finally, the PCILO procedure has recently been developed [6] and seems particularly well-suited to biochemical studies, essentially owing to its extreme

Table I. *All-valence and all-electrons calculations on fundamental biomolecules*

Molecules	Methods			
	EHT	CNDO	PCILO	Non-empirical
Purines and pyrimidines	X	X		X
Nucleosides and nucleotides	X		X	
Amino acids and peptides	X	X	X	X
Energy-rich phosphates	X			X
Pteridines				
Porphyrins	X			
Carotenes and retinals	X		X	
Quinones				
Flavins		X		
Nicotinamides				
Folic acid				
Pyridoxal				
Thiamine				
Steroids	X		X	
Saccharides	X		X	
Hydrogen bond	X	X		X
Pharmacological	X	X	X	

Table II. *Refined calculations on fundamental biomolecules. Representative references (see p. 100)*

Molecules	EHT	CNDO	PCILO	Non-empirical
Purines and pyrimidines	Pullman, A., 1968*, 1969*, 1970*. Pullman, B. and A., 1969. Rein et al., 1968.	Pullman, A., 1968*, 1969*, 1970*. Pullman, B. and A., 1969. Pullman, B., 1970. Fujita et al., 1969.		Mely and Pullman, A., 1969. Pullman, A., 1970*. Clémenti et al. 1969. Snyder et al. 1970.
Nucleosides and nucleotides	Jordan and Pullman, B., 1968.		Berthod and Pullman, B., 1971.	
Amino acids and peptides	Kier and George, 1970. Hoffmann and Imamura, 1969. Rossi, David and Schor, 1969. Yan et al., 1970. Govil, 1970, 1971. Colombetti and Petrongolo, 1971.	Pullman, A., and Berthod, H., 1968. Imamura, Fujita and Nagata, 1969. Yan et al., 1970. Kang et al., 1970.	Maigret, Pullman, B., and Dreyfus, 1970. Maigret, Pullman and Perahia, 1971. Pullman, Maigret and Perahia, 1970. Pullman, 1971.	Robb and Csizmadia, 1969. Dreyfus, Maigret and Pullman, A., 1970. Moffat, 1970.

Energy-rich phosphates	Collin, 1969. Boyd and Lipscomb, 1969. Boyd, 1970.		Boyd, 1970.
Porphyrins	Zerner and Gouterman, 1966.		
Carotenes and retinals	Pullman, B., Langlet and Berthod, 1969.		Langlet, Pullman,B., and Berthod, 1970.
Flavins		Song, 1969.	
Saccharides	Neely, 1970.		Giacomini and Pullman, B., 1970.
Steroids	Kier, 1968.		Caillet and Pullman, B., 1970.
Hydrogen bond		Pullman, A., and Berthod, 1968. Blizzard and Santry, 1969.	Dreyfus, Maigret and Pullman, A., 1970. Clémenti et al.,1971.
Pharmacological	Kier, 1970. Neely, 1970. Wohl, 1970. Yonezawa et al., 1969.	Andrews, 1969.	

* Review articles.

rapidity [7]. Its principle is the use of a perturbation procedure for computing extensive configuration interaction, starting with a set of localized orbitals, with all the approximations and parameters of CNDO/2. It goes beyond the SCF corresponding approximation and is much faster.

As to the all-electrons non-empirical computations, they are based on the Roothaan LCAO SCF-method [8], most of the applications in biochemistry using atomic basis sets of Gaussian functions [9].

The complete review of the present status of the result obtained in applying all the enumerated methods to biochemical compounds is beyond the scope of this paper. The reader can gain an idea of the extent of these developments by looking at Tables I and II where we have summarized the distribution, among the methods, of the main problems which have been dealt with, and the essential references in the field. In order to illustrate what can be gained by this sort of study, let us look at a few examples which we shall choose in connection with the most important biomolecules: the constituents of the nucleic acids and of proteins.

A. Fundamental Electronic Characteristics

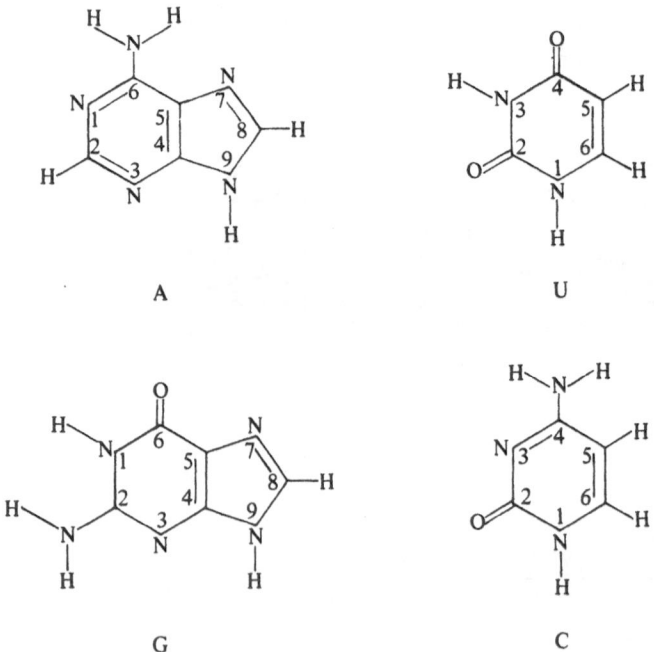

Fig. 1. The bases of the nucleic acids (Thymine is 5-methyluracil)

The two purines, adenine and guanine, and the three pyrimidines, cytosine, thymine and uracil (Fig. 1), are the fundamental components of the nucleic acids and, for this reason, have been extensively studied. Not only is the knowledge of their individual electronic structure of the utmost importance for the understanding of their biochemical reactivity, whereever they occur in the organism, but so it is also for the understanding or prediction of their various associating abilities, the best known of which are the base pairing by hydrogen bonds and the helical stacking of the base-pairs in DNA, although many other associations occur [10]. The evaluation of the intermolecular forces, responsible for the associations, requires the knowledge of a number of physico-chemical characteristics of the molecules involved: the electrostatic and inductive part of the Van der Waals-London forces requires polarizability and dipole moments (or charge distributions); the calculation of dispersion forces involves ionization potentials as well. All these quantities are very difficult to measure in such compounds. The experimental dipole moments of guanine and cytosine are still unknown and the ionization potentials have only been obtained very recently for A, C, T, and U[11].

Important predictions have been made for these quantities on the basis of π-electrons theory supplemented by simple σ considerations:

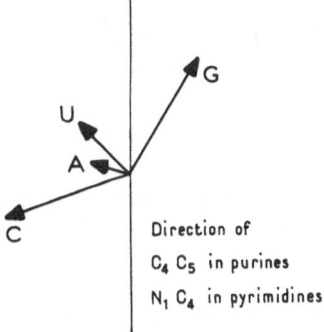

Fig. 2. Directions and relative values of the dipole moments predicted in the simple approximation (Pariser-Parr-Pople π moments added to Del Re-σ moments; details in Ref. [12])

a) The predictions on dipole moments are schematized in Fig. 2 [12]) The most important result is the relative ordering of the moments, predicting that guanine and cytosine should have a much larger moment than uracil, which in turn should have a larger moment than adenine. (This relative ordering was in fact already apparent in early Hückel evaluations [13,14].)

b) The increasing order of the π ionization potentials (or the decreasing order of the π electron-donor ability) is predicted to be [12]:

$$G < A < C < T < U$$

c) In these molecules, even when "lone pairs" are available, the π electrons should ionize first, oxygen lone-pairs coming next [15].

The quantities involved in these predictions are directly obtained in all-valence and all-electrons calculations. The results indicated for the dipole moments by EHT, CNDO/2 and non-empirical calculations are shown in Fig. 3 and Table III. It is rather striking that *all methods predict greater dipole moments for guanine and cytosine than for uracil, which in turn should have a slightly greater moment than adenine. What is still more striking is the similarity in the directions of the moments* predicted for the bases by all methods. Obviously the *overall* polarity of the molecules is well indicated by all procedures. A more

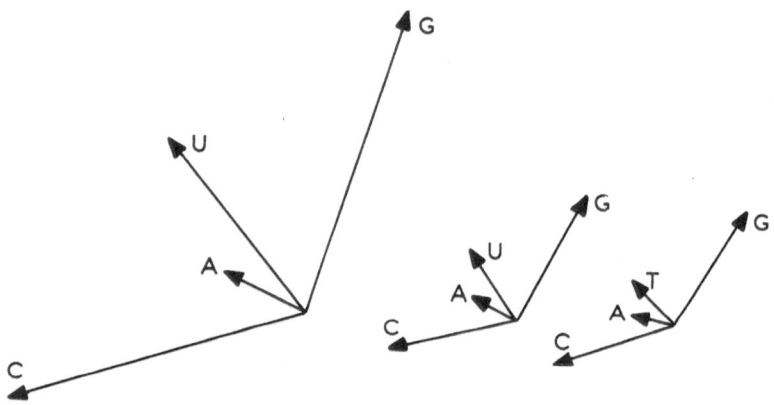

Fig. 3. Dipole moments in (left to right) EHT, CNDO/2, non-empirical calculations (same scale as Fig. 2)

detailed examination [16] has shown that the *trends* in the relative values and directions are embodied in the π components of the moments. The CNDO/2 method gives values in very close agreement both with the simpler representation and with the non-empirical method, and these values are in very satisfactory agreement with the experimental ones (available only for methylated uracil and adenine). It may, therefore, be expected that the corresponding values predicted for G and C are reliable.

As to the ionization potentials (when approximated by the negative of the molecular orbital energies), both EHT, CNDO/2 and non-empirical results agree on the fact that the highest occupied level is a π level[a]. Moreover, in A, C, G

---

[a] See Ref. [17] for a discussion of the iterative EHT results which may or may not invert the highest two levels according to the parameter values.

Table III. *Dipole moments of the nucleic bases (Debye units)*

	$\sigma + \pi$ [a]	EHT [b]	CNDO/2	Mély-Pullman	Clémenti	Experimental [c]
A	3.2	5.1	2.9	2.8	2.6	3.0
U [d]	3.6	12.3	4.6	4.2	3.3	3.9
G	6.8	16.6	7.3	–	6.9	–
C	7.2	16.5	7.6	6.8	6.4	–

[a] Ref. [14].
[b] Punctual charges.
[c] For 9-methyl adenine and 1,3-dimethyl-uracil.
[d] T for the non-empirical values.

Table IV. *Highest occupied π molecular orbital (ev), in all-valence electron methods*

	EHT	Iterative EHT	CNDO/2
G	11.84	10.31	9.06
A	11.95	10.43	10.08
C	12.50	10.39	10.78
U	12.71	11.08	11.88

Table V. *Highest occupied π molecular orbital (ev) in non-empirical SCF computations*

	Mély-Pullman	Clementi *et al.*
G		9.09
A	9.41	9.99
C	9.82	9.82 [a]
T	10.15	10.53

[a] Geometry different from that of all other calculations.

and U all-valence as well as all-electrons methods indicate that guanine should be the best π-electron donor and uracil the worst one (Tables IV and V) in agreement with the earlier predictions in the π electron approximation. A result more specific to the refined methods concerns the status of the lone pairs. Thus although atomic "lone pairs" do not exist as such (pure) in molecules, the analysis of the coefficients of the atomic orbitals in the molecular orbitals shows very neatly that a large amount of "lone-pair character" may be assigned to the highest σ levels in these heterocycles. Although deeper in energy than the highest π level, these "lone pairs" (or combined lone pairs) appear more ionizable than the other σ electrons, with oxygen lone pairs more ionizable than nitrogen lone pairs (see Ref. [16])). As already mentioned, theory has preceded experiment in this area. It is rewarding that the partial results now available entirely confirm the predictions [16].

It is thus seen that for the first property which we have considered, the dipole moment, quantitative predictions could be made with reasonable accuracy, even in the π-electrons era, with sufficient care and some training. The more elaborate computations have essentially brought confirmation of the main trends. For ionization potentials, the precision gained is more considerable, especially as concerns the relative locations of the π and "lone-pair" ionizations. (The separation of the highest π orbital and highest σ orbitals being over 1.5 eV in non-empirical computations, it is probable that correlation effects would not modify the relative ordering of the π and σ ionization potentials)

There is another area in which the all- or quasi all-electrons methods enable us to go much deeper than the previous approximations, namely the field of the *electronic distribution*. Thus, both π and σ *populations* are now directly attainable as well as *hybridization ratios,* which had previously to be assumed, both in the simple representation of the σ bonds and in the parameters of the π electron theory [12,14].

A detailed examination [16] of the results of the all- and all-valence electrons calculations has shown that they yield conclusions concerning the *overall* aspect of the *global* polarity which are nevertheless similar to those of the simple representation. Thus, the carbon atoms of the C_5-C_6 bond of the pyrimidines are of opposite polarity, with C_5 negative, whereas all carbons are globally positive (with the exception of C_5 of the purines in CNDO). The overall image of the NH_2 group shows a fair constancy. As to the hydrogen atoms, all procedures confirm an appreciable discharge for hydrogens bound to nitrogen (NH or NH_2), and generally a smaller discharge for hydrogen bound to carbon atoms. This last effect is smaller in CNDO, where it is even sometimes reversed, a feature which is characteristic of the procedure and disappears when deorthogonalization of the CNDO coefficients is performed [18,19].

The decomposition of the total charges into their σ and π components shows that oxygen atoms are both σ- and π-negative, (much more π than σ). All nitrogens are globally negative, but the pyridine-like nitrogens are both σ- and π-negative, the amino and pyrrole-like ones being σ-negative and π-positive. Some constancy in the positioning, by the π distribution, of the *most* reactive centers and in their relative ordering is observed in a number of cases: thus the most π-negative oxygen of all the bases is that of cytosine in all calculations. The order of decreasing positivity (increased delocalization) of the π lone pair of the glycosylic nitrogen is $G > A > C > U$. Another important constancy observed is the

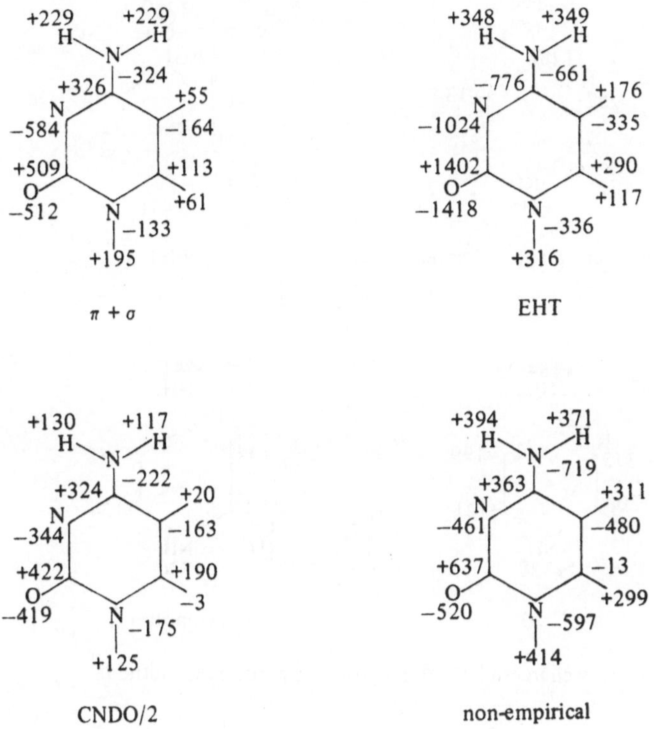

Fig. 4. Total charges (10^{-3} e) in cytosine by different methods

π-polarity of the C_5-C_6 bond in the pyrimidines with C_5 negatively charged and C_6 positively charged. Finally, the large negative π charge of N_3 of guanine appears in all results. However, the detailed ordering of the pyridine-like nitrogens with respect to their negative π-populations varies from one calculation to another.

As an illustration of the similarities and differences obtained, the total charges given by four procedures for cytosine are gathered in Fig. 4. Fig. 5 gives the corresponding π distribution. It is clear that if qualitative gross agreement is observed, some differences also appear. This is not surprising, particularly on account of the fact that "charges" in the different procedures are different quantities if only because of the different approximations involved. An interesting complement to the information given by the gross atomic charges is provided by the hybridization ratios around the different atoms, which can be obtained by inspection of the atomic orbital populations. The detailed analysis of the atomic-orbital populations for a number of molecules has shown [17] that in each procedure *the σ-population for a given atom in a given neighborhood is fairly constant regardless of the molecule considered, that it is very little depen-*

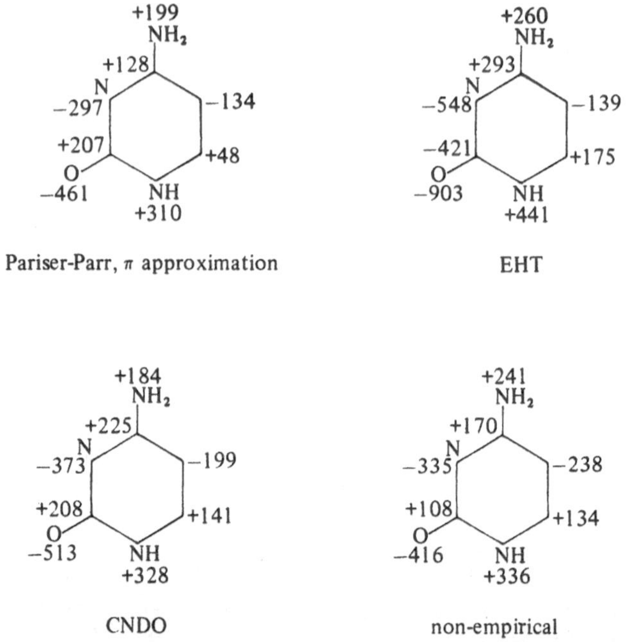

Fig. 5. π charges (10^{-3}e) in cytosine by different methods

Table VI. *"Hybridization" ratios in oxygen and nitrogen by different methods*

Atom	EHT		CNDO	Non-empirical
O	Carbonyl	$s^{1.8}p^{3.7}$	$s^{1.7}p^{3.2}$	$s^{2.0}p^{3.1-3.2}$
N	$-NH_2$	$s^{1.4}p^{2.5}$	$s^{1.2}p^{2.2}$	$s^{1.5-1.6}p^{2.4}$
N	$-\underset{\underset{H}{\mid}}{N}-$	$s^{1.4}p^{2.4}$	$s^{1.2}p^{2.3}$	$s^{1.5-1.6}p^{2.4}$
N	$=N-$	$s^{1.5}p^{3.0}$	$s^{1.4/1.5}p^{2.5/2.6}$	$s^{1.6-1.8}p^{2.4-2.6}$

dent of non-bonded atoms, and that it is but very slightly affected by the π-electron displacements. Typical ratios are given in Table VI. Note the difference observed for NH (or NH_2) nitrogens and pyridine-like nitrogens. Carbonyl oxygens display very constant ratios whatever the molecule. Another interesting characteristic is that they carry a very nearly pure p lone pair in the direction perpendicular to the CO bond (as shown by the populations in the p_x and p_y orbitals when CO is along the y axis, Table VII).

Table VII. *x and y populations for carbonyl oxygen in the three procedures (y along the C = O bond)*

	x	y
EHT	1.98	1.76
CNDO	1.92	1.27
Non-empirical	1.86	1.31

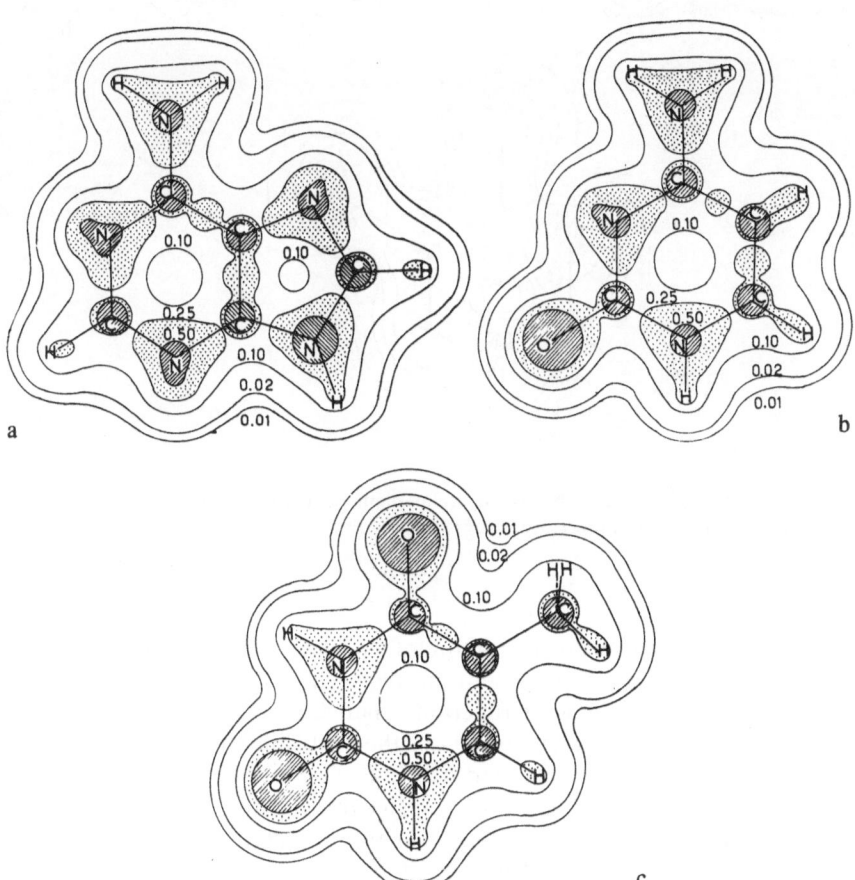

Fig. 6a–c. Global isodensity contours in the molecular plane. a) Adenine, b) cytosine, c) thymine (unit: e/a_0^3)

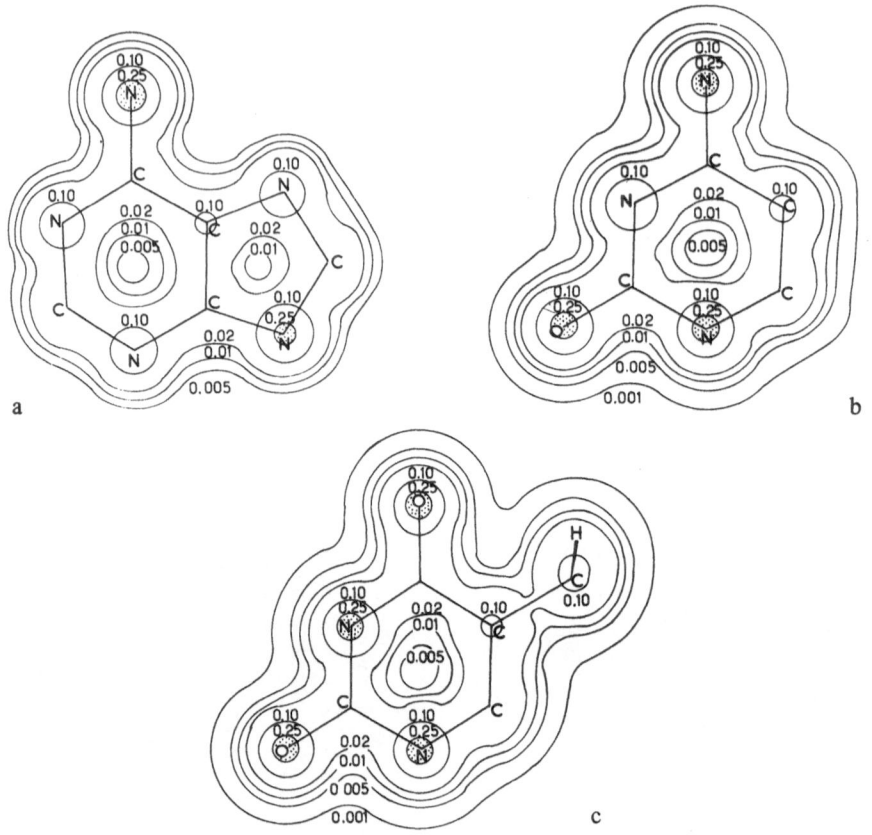

Fig. 7a–c. Isodensity contours above the molecular plane (as indicated in text).
a) Adenine, b) cytosine, c) thymine (unit: e/a_0^3)

Still more insight into the real structure of the electron distribution can be gained from the probability density contour maps which have the advantage of being a direct outcome of the wave function, whereas any population analysis represents, at best, a conventional division of the probability distribution in more or less well-defined atom-centered fractions. Fig. 6a, b, c and 7a, b, c give such contours obtained [20] for A, C and T, respectively in the molecular plane (x, y) and in a plane parallel to it, passing roughly through the position of the radial maximum of a nitrogen p_z atomic orbital.

The most striking feature of the σ contours is probably the constancy in the individual aspect of certain bonds, or atoms, or groups of atoms, in different environments: one can observe, in particular, the characteristic shape of the carbonyl group in thymine and cytosine, and the shape of the amino-group in adenine and cytosine. Such constancies are not limited to extra-cyclic substituents: inside the molecules, it can be seen that all the NH groups display a typical triangular form. Moreover, all the pyridine-type nitrogens are characterized by another triangular distribution in which the directional character of the lone pair is clearly visible. The carbon atoms appear globally very similar to one another, even outside the cycles, as in the methyl group of thymine. The carbon-carbon bonds of the cycles are well characterized by a "ganglion-type" aspect. The CH bonds are more or less "fat" with an overall similarity (the order of decreasing "fatness" of the CH bonds in cytosine and thymine follows the increasing order of the bond length adopted as input).

The localized character of the σ-cloud appears also in the relative symmetry of the hexacycle in adenine with respect to the C_2 C_5 axis, in spite of the substituent on C_6. A similar symmetry is seen in the cycle of thymine with respect to the same axis, in spite of the carbonyl on C_4.

Another outstanding feature of the σ-diagrams concerns the "lone pairs" in these molecules: the pyridine-like nitrogens show a very distinct elongation of their electronic cloud towards the outside of the rings in a direction bisecting the external angle of the single bonds. On the other hand, in the case of the carbonyl oxygens, no directional character appears in the electron distribution and the non-bonding electrons are entirely buried inside a nearly spherical distribution. This last result is in agreement with the conclusions obtained by comparison of X-rays and neutron diffraction data on carbonyl oxygens [21].

This non-directional character of the *global* electron cloud around the carbonyl oxygen should not be taken as erasing completely the old notion of carbonyl lone pairs. The total density considered is a sum of molecular orbital contributions: among the doubly occupied molecular orbitals obtained in the SCF calculations, two are essentially localized in the neighbourhood of the oxygen, their probability distribution having the shape indicated on Fig. 8 (a and b). These particular components of the total density show up in some properties where their particular lability or polarizability appear, but there is no doubt that the simplified representation of the sticking-out carbonyl lone pairs should not be taken too literally, especially as concerns global density [22].

As to the π-electron density, the map of adenine shows very neatly the distinction between two types of nitrogens, the "pyrrole- or amino-type" and the "pyridine-type". In the molecular field, the first ones lose part of their pair of π electrons while the second type gains π-charge over its original one-electron contribution. The final distribution shows more π-density left on the first kind than gained on the second. This agrees with the relative values found for the gross atomic populations. Moreover, in the three molecules, π-electrons of NH_2 delocalize less than those of NH, in agreement with both the classical concepts and with the corresponding gross populations.

The carbon atoms of the cycles appear very little charged in π electrons

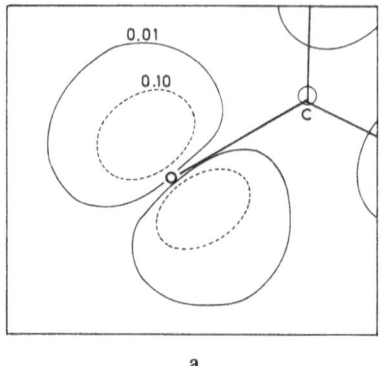

a

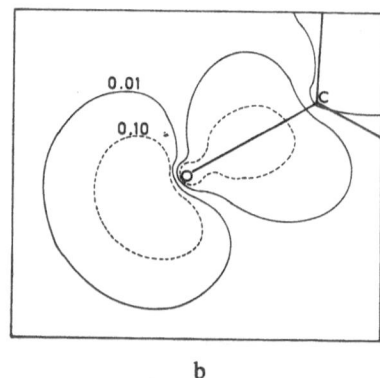

b

Fig. 8a and b. Isodensity contours (e/a_0^3) for the oxygen "lone-pair" doubly occupied SCF orbitals in cytosine. a) Highest filled σ orbital, b) third σ orbital

with the exception of C_5 in the three molecules with a clearly larger extension of the high density region in cytosine. This result thus confirms the strong π-polarity of this particular atom, mentioned earlier.

The carbonyl oxygens appear with a strong π-polarity.

Although the fine details of the electron density maps evidently differ according to the atomic orbital basis used in the SCF calculation, the conclusions which we have just discussed do not change with a change in the basis set [20].

B. New Approach to Chemical Reactivity: Molecular Isopotential Maps

It has been customary in the past to utilize, in the study of reactivity, indices based on the knowledge of the molecular electronic structure. There is no difficulty, in principle, in defining in all-valence or all-electrons methods reactivity indices, whether "static" or "dynamic" [23], similar to those used in the π electron-theory. But perhaps more fruitful is the exploration of new approaches developed under the impulse of the new methods. One seemingly very promising line is the study of the *molecular potential* seen by an approaching reagent [24]: the knowledge of a molecular wave function allows the accurate calculation of the electrostatic potential $V(r_i)$ created in the neighboring space by the nuclear charges and the electronic distribution[a]. The interaction energy

---

[a] For a first-order density function $\rho(1)$, the average value of the potential $V(r_i)$ at a given point i of space is [25]

$$V(r_i) = - \int \frac{\rho(1)}{r_{1i}} \, d\tau_1 + \sum_\alpha \frac{Z_\alpha}{r_{\alpha i}}$$

where Z_α is the nuclear charge of nucleus α.

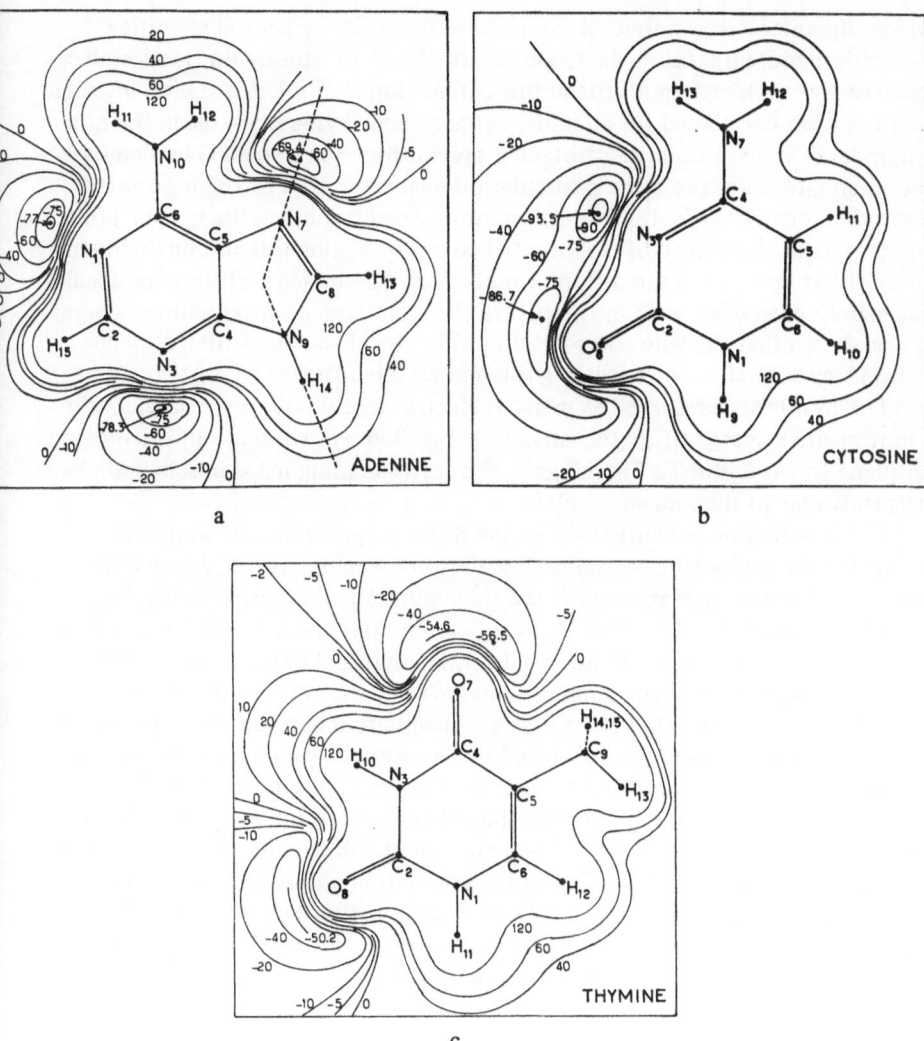

Fig. 9a–c. Isopotential maps for an approaching unit positive charge in the molecular plane.
a) Adenine; b) cytosine; c) thymine (unit: kcal/mole)

between the molecular distribution (considered unperturbed) and an external
point charge q placed at point i is $qV(r_i)$, which is rigorously the first-order
perturbation energy of the molecule in the field of the charge q. Its very definition makes it an appropriate index for studying chemical reactivity, at least
in the early phase of approach of an external reagent. Taking q as a *unit positive*

61

charge the interaction potential can be used for studying proton affinities (basicities) and, hopefully, electrophilic attacks. Early studies on small molecules have given interesting results in this connection [24,25-27]. A detailed investigation has been performed for adenine, cytosine and thymine [28], using the nonempirical SCF wave functions obtained previously [29] in a small Gaussian basis set. Such molecules present a particular interest since each of them possesses more than one possible site for protonation or electrophilic attack, thus providing a good test of the ability of isopotential curves to distinguish among different positions of attack, or among different molecules. The basicity of these molecules has mainly been dealt with in the past in the π electron approximation, where it was shown that the sole consideration of the local charge of the atoms involved was not sufficient to distinguish among the different sites [30].

The results are presented as maps of electrostatic interaction energies with a unit positive charge (isopotential maps). Fig. 9 gives such maps in the molecular plane for A, C and T. Fig. 10 gives the corresponding maps in selected planes perpendicular to the molecular plane.

Examination of the curves shows the distinct appearance of well-defined characteristic regions of attraction for the approaching proton. *In adenine* three such regions appear towards the three pyridine-like nitrogens N_3, N_1, N_7, the regions around N_1 and N_3 being very similar in shape and depth, whereas the N_7 region is narrower and less deep. The NH and CH regions as well as the inplane approach to the amino group are repulsive. The perpendicular section of Fig. 10a shows that there exists a secondary minimum above (and below) the amino nitrogen. The shape of the large attractive region covering the whole N_{10} C_6 N_3 area indicates that, when both are present, the basicity of a pyridine-like nitrogen is greater than that of an NH_2 group, in agreement with well-known chemical facts [31,32]. The experimental data concerning electrophilic attack on adenine indicate that both protonation and alkylation involve the ring nitrogens. Among them, the preferred regions of attack are definitely N_1 and N_3, a distinction among the two positions being difficult: results on the base itself indicate N_1 as the first protonation site [33,34] but N_3 was found to be more reactive than N_1 towards alkylating agents [35]. In nucleotides, nucleosides, and in the nucleic acids, N_1 seems to be the first protonation site [36], alkylation occurring at both N_1 and N_3 [37]. In RNA, N_1 is the principal minor site of alkylation (the major site is N_7 of guanine) while this site shifts to N_3 in DNA (where N_1 is involved in hydrogen-bonding)[38] [a].

Cytosine presents a new situation due to the presence of a carbonyl group adjacent to a pyridine-like region for electrophilic agents appears on this side of the molecule, with two deep minima, one located in the direction of the nitrogen lone pair, the other at an angle of 55° from the C=O bond. This last location is quite interesting in view of the previously mentioned fact, that the *global*

---

[a] See Ref. [39] for a very recent summary and discussion.

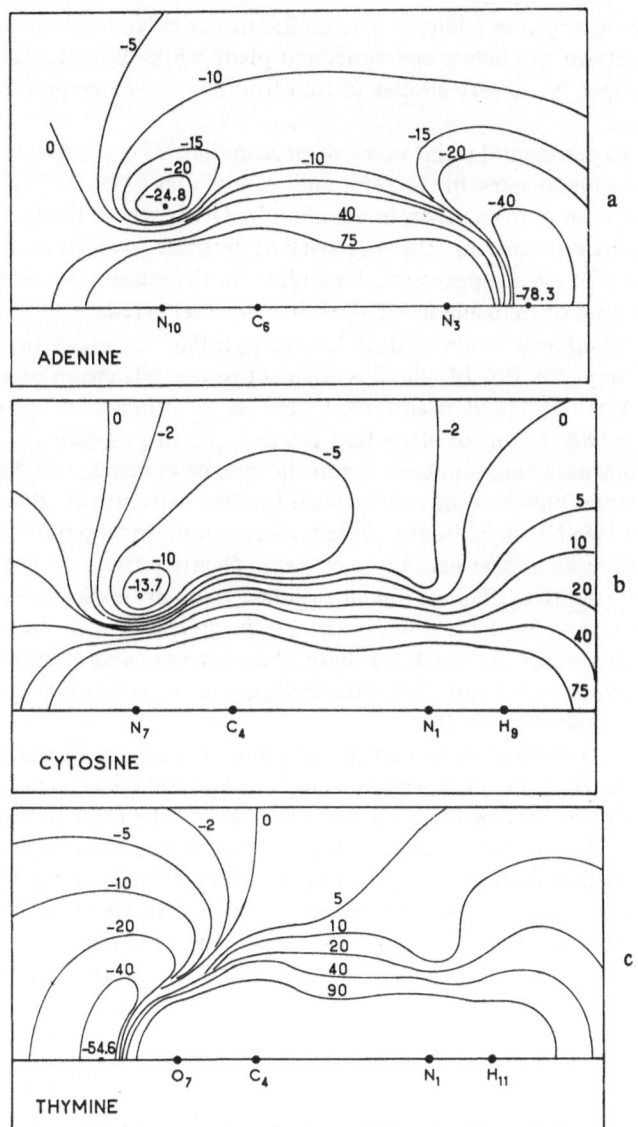

Fig. 10a–c. Isopotential maps in planes perpendicular to the molecular planes; a–c) as indicated. Same unit as in Fig. 9

isodensity curves show no directional properties around the carbonyls[a]. The potential well is deeper for N_3 than for O_8 and the minimum for N_3 is deeper

---

[a] A discussion of the location of the potential minimum and its relation to the hidden "lone-pair" orbitals can be found in Ref. [28].

than for either nitrogen in adenine. The amino nitrogen N_7 has two equivalent small minima above and below the molecular plane while the potential around the imino nitrogen N_1 is very similar to that found in the corresponding N_9H zone of adenine.

From the experimental point of view, protonation of cytosine, its nucleoside and nucleotide, occurs on N_3 [33,40] and so does alkylation [38,41], both reactions being easier in cytosine than in adenine. In DNA, where the N_3 position is involved in hydrogen bonding, the reactivity of cytosine towards methylation is much reduced or even suppressed, depending on the reagent [38]. A protonation of the oxygen of cytosine in DNA [42] has been reported.

As far as *thymine* is concerned, it has no pyridine-like nitrogen, but has two carbonyl oxygens, one of which is adjacent to one NH group on each side; the other has one NH neighbor and one CCH_3 on the other side. The potential curves indicate two regions of attraction towards the two oxygen atoms, the rest of the molecule being repulsive. As in the case of cytosine, the directionality of the attraction is strongly influenced by the environment: the oxygen O_8, which has two NH neighbours, presents one symmetrical potential well, whereas two minima appear near O_7, one being clearly pushed away from the NH region. The depth of the minima in thymine is much smaller than that of the corresponding wells in cytosine, in complete agreement with the fact that thymine and thymidine are much less basic than cytosine and cytidine [38] and that neutral thymine does not undergo alkylation in the conditions under which the other bases react [34,35,38,41].

This discussion shows clearly that the positions and depths of the potential wells are undoubtly connected with the ease of electrophilic attack. Even though the electrostatic energy considered is but a fraction of the total energy of interaction of the molecule with an approaching proton, it appears as a determining element. Although polarization and charge transfer effects are neglected in this representation, it seems – in the case studied – as if they are either less important than sometimes supposed or have their maxima in the same directions as those given by the consideration of the electrostatic energy alone. Investigation of this question is under way, as well as studies of the influence of the accuracy of the wave function. Studies using the CNDO method have also been performed [43] and have proven extremely encouraging, showing in particular the fundamental qualitative difference in the molecular electrostatic potential created in the neighbouring space by adenine and guanine. Thus, Fig. 11 a shows the characteristic potential wells facing N_1 and N_3 of adenine, with the smaller region near N_7, while the diagram of guanine (Fig. 11 b) presents its strong attractive region in the neighbourhood of N_7 and of the carbonyl oxygen, in very striking agreement with the experimental facts concerning protonation and alkylation of these molecules [37,38,44,45].

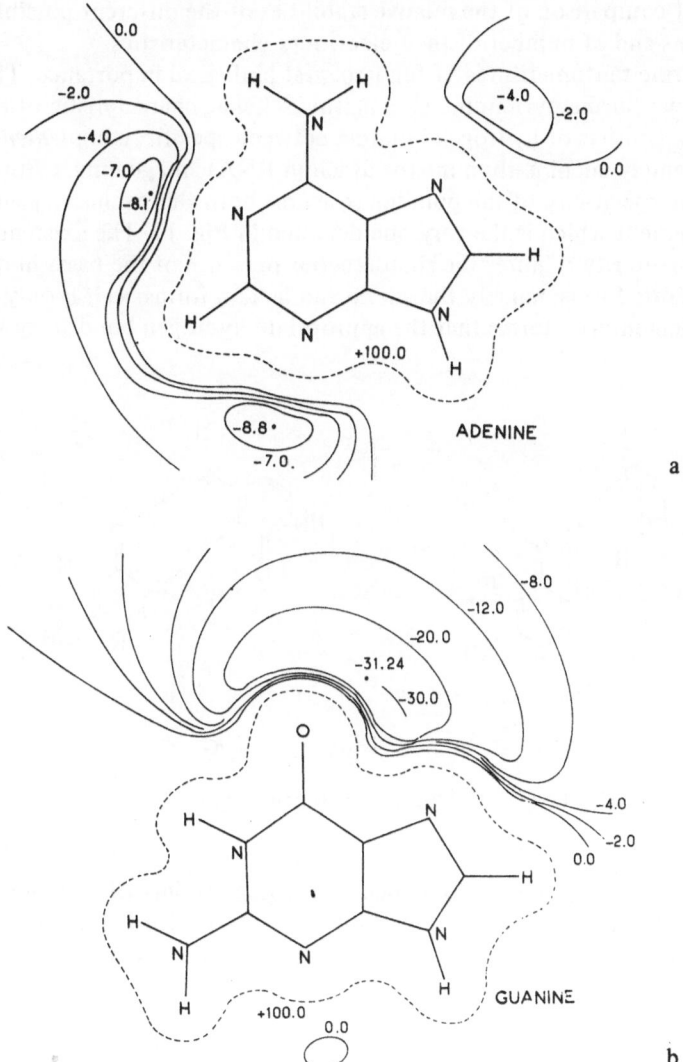

Fig. 11a and b. Isopotential maps (CNDO/2 approximation) in the planes of a) adenine, b) guanine

C. Tautomerism

A capital problem in the domain of purines, upon the study of which the introduction of the all-valence electrons methods had a particularly striking impact, is their *tautomerism:* indeed, the calculation of the *total* molecular energies en-

65

ables a direct comparison of the relative stabilities of the different possible tautomeric forms and of number of their electronic characteristics.

Now, purine tautomerism is of fundamental biological importance. Thus, the well-known purine-pyrimidine base pairing scheme, characteristic of the nucleic acids, consists of hydrogen bonding between specific, *complementary* base pairs, namely adenine-thymine (or uracil in RNA) and guanine-cytosine (Fig. 12). The specificity of the bonding concerns both this exclusivity and its steric arrangement which is the very one depicted in Fig. 12. The existence of this complementarity requires the simultaneous presence of the bases in definite tautomeric forms, namely the amine and lactam forms, as it is only with such "complementary" forms that the appropriate hydrogen bonds may be formed.

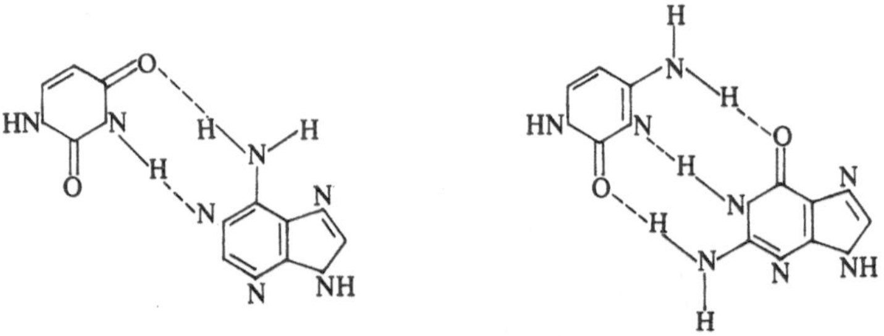

A – T   G – C

Fig. 12. The Watson-Crick base pairing in DNA

There is abundant evidence, obtained with a large variety of experimental techniques, that these are the common tautomeric forms for the majority of biological purines and in particular for those involved in the nucleic acids, both in the solid state and in solution.

Should one of the bases occur in a rare tautomeric form, say imine for adenine or cytosine and lactim for guanine or uracil, *miscouplings* may occur. This is illustrated in Fig. 13, which shows how cytosine in its rare *imine* form would be able to hydrogen-bond to adenine in its common form (and *vice versa*) and guanine in its rare lactim form could couple with thymine in its usual form (and *vice versa*). As a result of such *miscouplings* the original order of the arrangement of successive base-pairs along the axis of the nucleic acid would be modified and the modification perpetuated during DNA replication. The order of the complementary base-pairs along the axis of DNA being most probably responsible for the genetic code of the species, any perturbation of this order represents by definition *a mutation*.

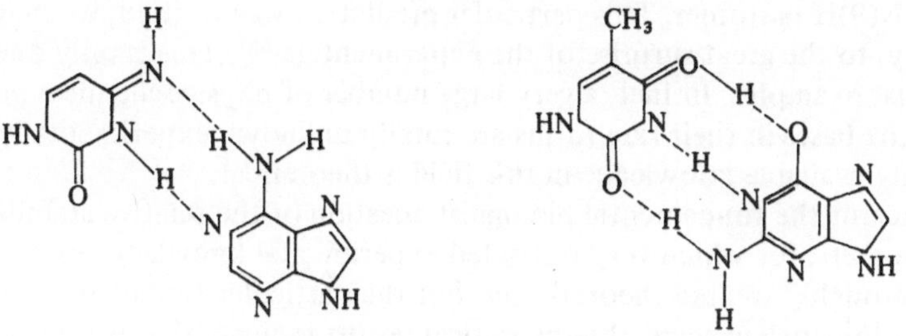

Cytosine (rare imino form)-
adenine (normal form)

Guanine (rare enol form)-
thymine (usual form)

Fig. 13. Examples of miscoupling of the bases of the nucleic acids

A fundamental importance therefore attaches to the study of the relative stabilities and electronic characteristics of the different possible tautomers of purine and pyrimidine bases. These are in fact investigated, as already said, by a number of experimental techniques and have also been investigated extensively by quantum-mechanical calculations, mostly by the CNDO/2 procedure. (For a general account see Ref. [46] and [47].) The results of these last computations compete quite advantageously with the results of experimental studies, in fact, they frequently anticipate the latter by a number of predictions. Among these, one of the most useful concerns the variation of the dipole moment with tautomerization which, owing to the theoretical predictions, has become an analytical tool for the identification of the tautomers. Particularly striking in this field were the predictions concerning the prototropic tautomerism between the nitrogen atoms N_7 and N_9 of purines:

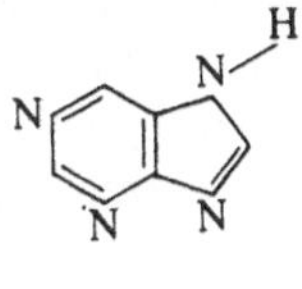

N(7)H tautomer of purine

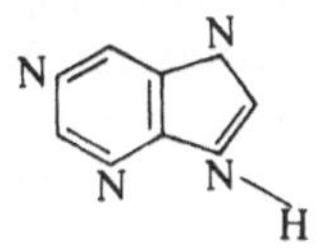

N(9)H tautomer of purine

It was the merit of the theory to have attracted attention to the large variations (2 to 5 Debye units) of dipole moments possibly associated with such tautomerism and to the specific directions of the variation following the nature of the purine. Thus the prediction was that, while the dipole moment of the N(7)H tautomer of purine should be greater than that of its N(9)H tautomer, the dipole moment of the N(7)H tautomer of xanthine should be smaller than that of its N(9)H tautomer. This particular prediction was verified, we may confidently say, to the great surprise of the experimenters [48]. This is only one of many similar examples. In fact, a very large number of physicochemical properties of the bases in their rare forms are totally unknown experimentally, and the only available knowledge in this field is theoretical.

As concerns the fundamental biological question of the relative stabilities of the tautomers, for which very restricted experimental knowledge exists, it has been throughly treated theoretically. For the particular case of the constituents of the nucleic acids, the theoretical results indicate that it is guanine and cytosine which should have the greatest tendancy to exist in their respective (lactim and imine) rare forms. These are therefore the bases most likely to be involved in spontaneous mutations, in so far, of course, as tautomerization may be considered as a cause of such mutations. The transformation $G - C \longrightarrow A - T$ should thus be more frequent than the reverse one.

It may be interesting to indicate that the fact that the $G - C$ pairs constitute the unstable part of the genome and that they mutate spontaneously more frequently than the $A - T$ pairs has indeed been reported in a number of publications. One cannot and should not conclude from this situation that miscouplings through tautomerization of bases *are* the principal cause of spontaneous mutations, because these are definitely known to be due to a large series of causes, the relative importance of which is difficult to ascertain. The concordance between the theoretical predictions and the experimental facts is, however, worth stressing, as any discordance would certainly have been picked up.

III. Calculations on the Conformation of Biomolecules

The second fundamental aspect of the recent developments that we shall examine is the conformational one. As is well known, the activity of biological molecules and, in particular, of biopolymers is frequently strongly dependent upon their conformation. The understanding of the factors governing conformational stability of biomolecules and the evalution of the preferred conformers is therefore of utmost interest for the further development and broadening of quantum biochemistry. It is one of the major contributions of the all-valence electrons methods to have made this development feasible.

The need for such a promotion of quantum mechanical studies appears the more necessary as there has been during the last 5 years an extremely striking development of what may be called "empirical" studies in this field. These consist in partitioning the potential energy of the system into several discrete contributions, such as non-bonded and electrostatic interactions, barriers to internal rotations, hydrogen bonding, etc., which are then evaluated with the

Table VIII. *Calculations on conformation of biological molecules*

Molecules	Methods					
	Empirical	Semi-quantum	EHT	CNDO	PCILO	Ab initio
Amino acid residues and polypeptides	X	X	X	X	X	
Pyrimidines (dihydro and T dimers)			X			
Constituents of nucleic acids, in particular nucleosides and nucleotides	X	X	X		X	
Mono, oligo and polysaccharides	X		X		X	
Polyphosphates			X			
Carotenes and retinals		X	X		X	
Steroids			X		X	
Coenzymes					X	
Hydrogen bond	X			X		X

help of empirical formulae deduced from studies on model compounds of low molecular weight (for general reviews of such works, see [49-53]).

Praiseworthy as such attempts are, they suffer from two obvious drawbacks. In the first place, whatever the practical justification for the partitioning of the total potential energy into a series of components, the procedure necessarily involves an element of arbitrariness and, possibly, incompleteness. Secondly, the fundamental formulae and parameters used to define the various components are far from being well established and differ, often appreciably, from one author to another. A mere rigorous deal may be expected from a quantum mechanical approach which is able to evaluate the total molecular energy corresponding to any given configuration of the constituent atoms and thus to choose the preferred ones.

Table VIII indicates schematically the principal groups of biomolecules whose conformational problems have been dealt with quantum-mechanically (and for comparison, also those which have been investigated "empirically"). Table IX lists the principal references to these works.

We shall describe rapidly some of the principal results obtained.

Table IX. *Calculations on conformations of biological and pharma-*

Molecules	Methods	
	Empirical	EHT
Amino-acid residues and polypeptides	Ramachandran and Sasisekharan, 1968 Scheraga, 1968 Ramachandran, 1968, 1969. Flory, 1969. Liquori, 1969. Popov and co-workers, 1968. Ponnuswamy, 1970.	Hoffmann and Imamura, 1969; Kier and George, 1970; Rossi, David and Schor, 1969, 1970. Govil, 1970, 1971.
Constituents of nucleic acid in particular nucleosides and nucleotides	Haschemeyer and Rich, 1967. Sasisekharan and co-workers, 1967, 1969.	Jordan and Pullman, B., 1968. Govil and Suran, 1971.
Oligo- and polysaccharides	Rao and coworkers, 1967, 1969. Sudararajan and Rao, 1969, 1968. Rees and Skerett, 1968, Rees and Scott, 1971. Goebel, Dimpfl and Brant, 1971.	
Polyphosphates		Collin, 1969. Boyd and Lipscomb, 1969.
Carotenes and retinals		B. Pullman, Langlet and Berthod, 1969.
Steroids		Kier, 1968.
Hydrogen bond	Ramachandran and coworkers, 1966. Liquori, 1969.	

cological molecules (References see p. 102)

	Methods	
CNDO/2	PCILO	Non empirical (ab initio)
Imamura *et al.*, 1969; Bossa *et al.*, 1971; Yan *et al.*, 1970.	Maigret, Pullman, A., and Dreyfus, 1970. Maigret, Pullman, B., and Perahia, 1970. Pullman, Maigret, and Perahia, 1970. Maigret, Perahia, Pullman, B., 1970. Pullman, B., 1971.	
	Berthod and Pullman, A., 1971.	
	Langlet, Pullman and Berthod, 1970.	
	Caillet and Pullman, 1970.	
Momany *et al.*, 1970. Pullman,A., and Berthod, 1968.		Dreyfus and Pullman, 1970.

A. Pullman

A. The Conformation of Dipeptides: Prelude to the Study of the Conformation of Polypeptides and Proteins

In biochemistry, the most important conformational problems undoubtedly concern proteins, whose enzymic activity frequently depends on assuming the appropriate spatial configuration. Most of the afore-mentioned "empirical" studies have been devoted to this group of substances and to their constituents and analogs (di-, oligo-, and polypeptides) and precise conventions have been established for presenting the subject [54]. Their essential aspects, virtually self-explanatory, are illustrated in Fig. 14 on the example of a dipeptide. The peptide units being considered as planar ($\omega = O$) and *trans*, the conformation of the backbone of the chain is described by specifying the values of the dihedral angles Φ and Ψ which, by convention, are equal to zero for the fully extended form and are measured in a clockwise direction from that conformation. The conformation of the atoms in the side chain may be denoted by a series of dihedral angles χ^1, χ^2 etc., corresponding to rotations about the bonds $C^\alpha - C^\beta$, $C^\beta - C^\gamma$ etc., from a standard conformation (taken to be that in which the two atoms of the bond, the heavy atom before and the heavy atom behind, are all in a plane and the last two are *cis* with respect to each other).

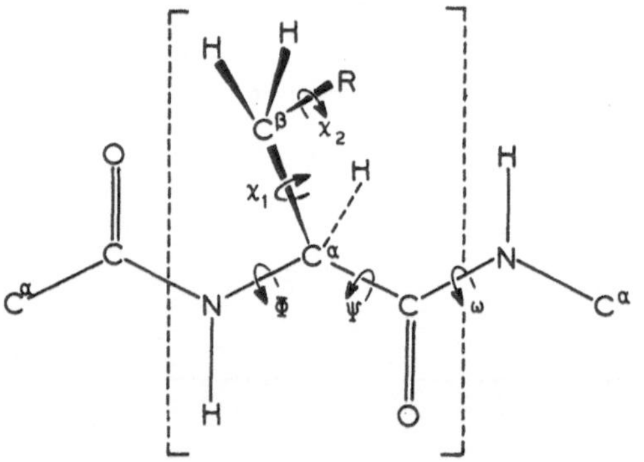

Fig. 14. A dipeptide and standard conventions for the study of polypeptide conformations
⌐ ¬ Limits of a residue
⌊ ⌋

It is fully acknowledged today that the determination of the "conformational maps" (values of potential energy as a function of Φ and Ψ) for the different amino acid residues of proteins may represent an essential step on the road leading towards the theoretical elucidation and prediction of protein conformations. It is therefore natural that their study should have been at the center of "empirical" research and was also the first objective of the quantum mechanical evaluations. The evaluations have been carried out extensively, in particular by

72

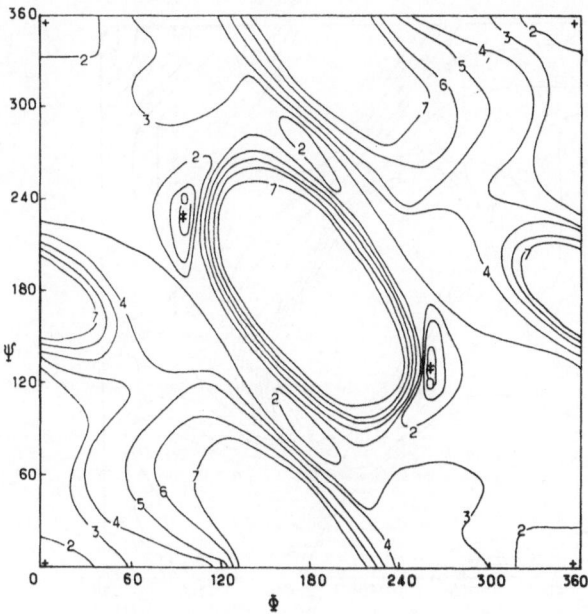

Fig. 15. Conformation map of the glycyl residue (PCILO calculations). Isoenergy curves in kcal/mole with respect to the global minimum ‡. + Local minima

the PCILO method. Specifically they represent the search for the allowed, and among those of the preferred conformations of dipeptides of the general formula I:

$$
\begin{array}{c}
\overset{\displaystyle H}{|}\ \ \overset{\displaystyle H}{|}\ \ \overset{\displaystyle O}{\|}\\
C_3C-C-N \rightarrow C \rightarrow C-N-CH_3\\
\overset{\displaystyle \|}{O}\ \ \ \ \overset{\Phi\ |\ \Psi}{R}\ \ \ \ \overset{\displaystyle |}{H}
\end{array}
$$

I

referred to in abbreviation as glycyl-L-glycine for R = H, glycyl-L-alanine for R = CH$_3$ etc., the corresponding conformational maps being frequently designated also as referring to the glycyl, alanyl, etc. residues.

Let us look at some typical results and compare them on the one hand with the results of the "empirical computations" and on the other with the results of the X-ray studies on the conformation of the amino acid residues in the crystals of globular proteins. Fig. 15 and 16 represent, by way of example, the conformational energy maps (isoenergy curves as a function of the rotations about the Φ and Ψ angles) for the two simplest residues, glycyl (GLY)

73

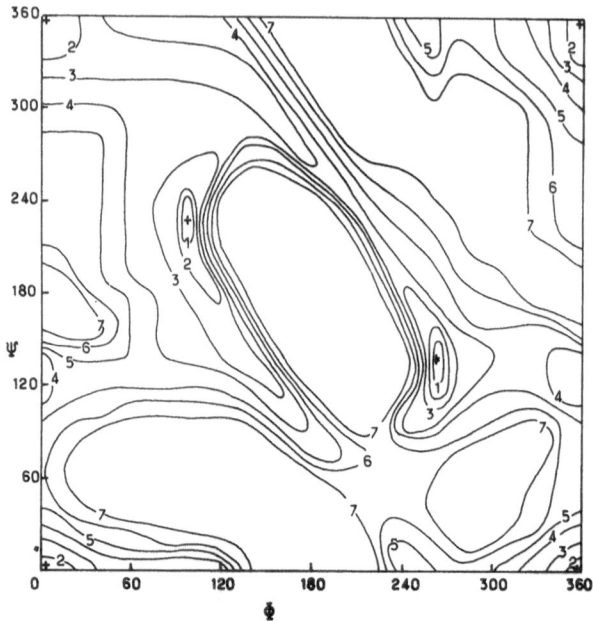

Fig. 16. Conformational map of the alanyl residue (PCILO calculations). Isoenergy curves in kcal/mole with respect to the global minimum ‡. + Local minima

and alanyl (ALA). Fig. 17 and 18 indicate in more detail the comparison with the data from empirical computations and from experiment. Thus they contain: 1) a typical contour of the zone of allowed conformations obtained by empirical computations; 2) the contour of the stable zone allowed by the quantum-mechanical computation within the limit of 6 kcal/mole with respect to the global minimum, the limit being compatible with conformational transformations without chemical ones; 3) the experimental conformations of these residues as determined by crystallographic studies on eight globular proteins: lysozyme, myoglobin, α-lactalbumin, ribonuclease-S, carboxypeptidase A, α-chymotrypsin, erythrocruorin and subtilisin-BPN' and for GLY in a number of small molecules.

In these conditions the confrontation leads immediately to two essential conclusions:

1) It is seen that in both cases the quantum-mechanical calculations impose fewer restrictions on the allowed or preferred conformational space than do the empirical ones. Thus, while for GLY this space is made up in the empirical computations by four separate zones A, B, C and D, these zones communicate between themselves in the PCILO calculations. In the case of ALA, the PCILO calculations substantially increase the limits of this space, with re-

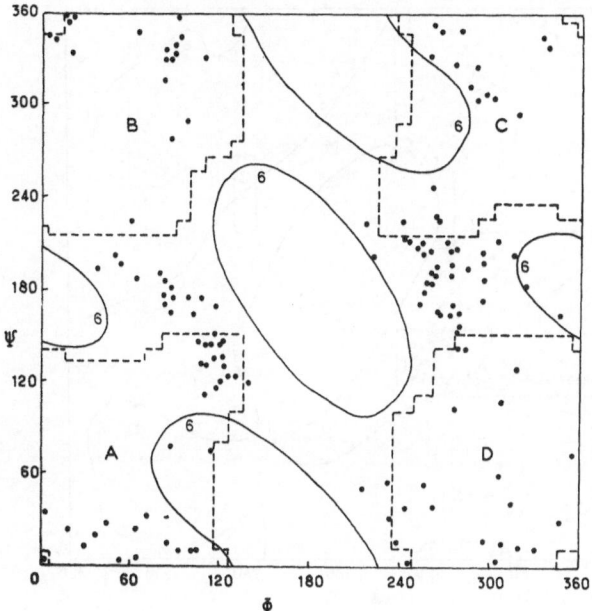

Fig. 17. Conformational map of the glycyl residue

- - - - - Typical limits of empirical computations
——— Limits of PCILO calculations within 6 kcal/mole above above the global minimum
■ Conformations of the GLY residues in globular proteins

spect to the empirical ones through the adjunction of new regions: region E, the corridors N and M and the extension of A towards P.

2) The agreement between theory and experiment is much better with the quantum-mechanical calculations than with the empirical ones, many representative points lying in the space forbidden by the latter but allowed by the former. Particularly important from that point of view are the regions comprised between A and B or C and D for GLY and the region N for ALA.

Similar results have by now been obtained for practically all the aminoacid residues of proteins and they all lead to similar individual conclusions.

They indicate also, however, some general features which enable a substantial extension of the application of the procedure. Thus, one of the basic conclusions which may be drawn from these results is that, as already considered by some of the authors engaged in empirical computations, the general allowed conformational space, within a fixed limit above the individual deepest minima, is similar in all these residues and similar to the general contour obtained for the alanyl residue. Naturally, there are differences among the individual residues but they all more or less conform to the pattern obtained for alanyl.

75

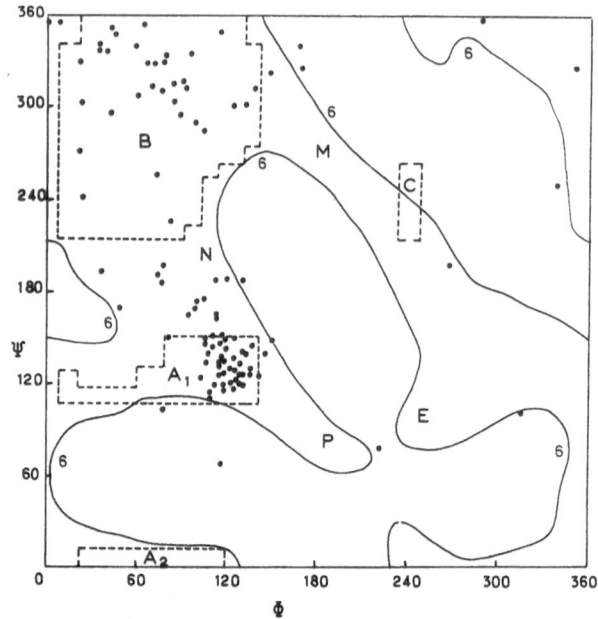

Fig. 18. Conformational map of the alanyl residue
----- Typical limits of empirical computations
——— Limits of PCILO calculations within 6 kcal/mole above above the global minimum
Conformations of the ALA residues in globular proteins

What essentially distinguishes these residues from alanyl (and among them-selves) is the fine structure of the conformational space and the location of the different energy minima.

Under these conditions it seems reasonable to assume that the conforma-tional map of alanyl residue may be considered as representing, at least at a first approximation, the conformational map of all C^β containing residues, and hence of all residues with the exception of glycine. In fact another exception is offered by proline which, because of the special feature of the pyrrolidine ring, represents a special case [55].

In order to check the validity of this proposition, we have compared the theoretical predictions based on this assumption with the available experimen-tal data on all the amino-acid residues in a number of experimentally deter-mined proteins [56]. Fig. 19, 20 and 21 illustrate this comparison for the pro-teins lysozyme, myoglobin and α-chymotrypsin, respectively.

The results of this comparison are most striking:

1) The 6 kcal/mole quantum-mechanical contour of the alanyl residue ac-

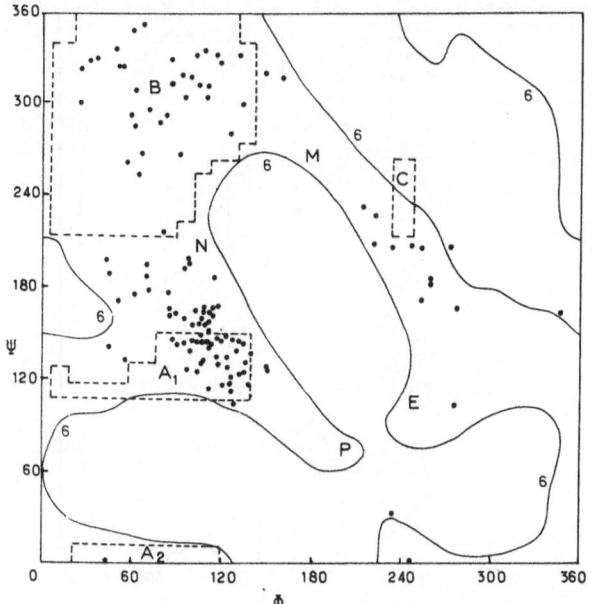

Fig. 19. The general significance of the conformational map for the alanyl residue

----- Allowed limits of the "hard sphere" approximation

—— Limits of the quantum-mechanical computations within 6 kcal/mole above the deepest minimum (‡)

■ Conformations of all the amino acid residues in lysozyme

counts excellently for the position of practically all the representative points in the three Figs. 19, 20 and 21.

2) The agreement between theory and experiment is appreciably better with the results of the quantum-mechanical computations than with those of the empirical ones. In particular, one may note the obvious significance in this respect of the regions E, M, P and N of the quantum-mechanical computations. Although the density of the representative points is certainly smaller in these regions than in the A_1 and B regions, their omission by the empirical computations represents a significant defect.

A large number of other problems have been dealt with and a substantial amount of other significant results obtained in this field. Lack of space prevents their description here. They may be found by consulting the original references indicated in Table IX.

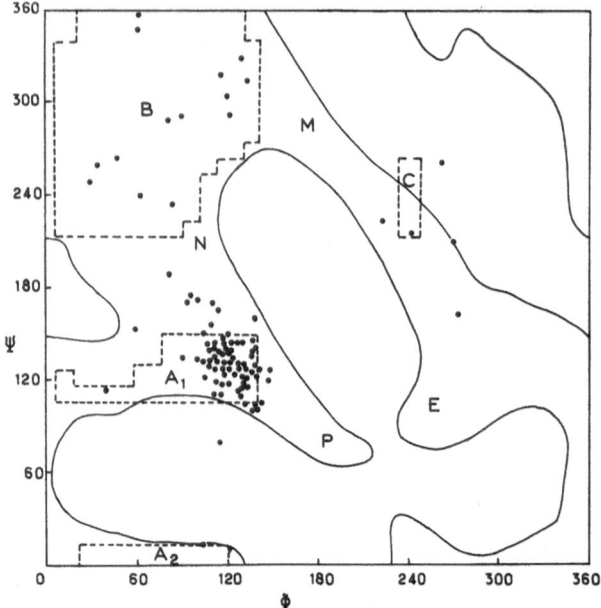

Fig. 20. Same as Fig. 19, for myoglobin

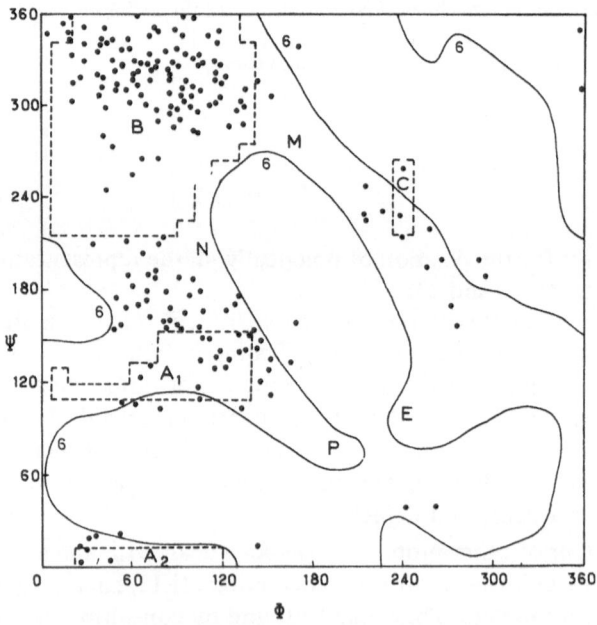

Fig. 21. Same as Fig. 19, for α-chymotrypsin

B. Nucleic Acid Constituents

Although potentially rich in possibilities, the conformational problems connected with the nucleic acids, their constituents and analogs, have received much less attention from the theoreticians than those of proteins. Active work is, however, starting in this field too. The fundamental degrees of freedom to be taken into consideration are numerous, as can be seen from the image of the principal torsion angles indicated for a nucleotide unit in Fig. 22. The quantum-mechanical studies have up till now centered essentially on the question of the *syn* or *anti* conformation of the purine and pyrimidine nucleosides and nucleotides. The problem corresponds to the study of the torsion angle χ of Fig. 22, generally designated χ_{CN}. This torsional angle about the glycosidic

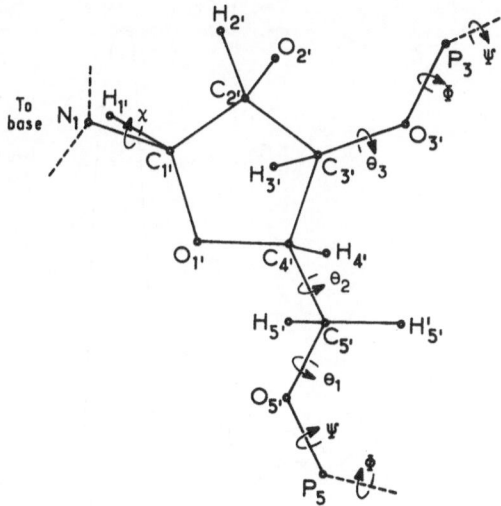

Fig. 22. The nucleotide unit

bond C(1')–N defining the relative orientation of the base with respect to the sugar, is the angle of rotation of bond N(9)–C(8) in the purines (or of the bond N(1)–C(6) in the pyrimidines) about the C(1')–N(9) (or C(1')–N(1)) glycosidic bond relative to the C(1')–O(1') bond of the ribose. The zero value of the angle is obtained when the N(9)–C(8) bond of the purines (or the N(1)–C(6) bond of the pyrimidines) is *cis*-planar to the C(1')–O(1') bond of the sugar with respect to the rotation about the glycosidic bond. Positive rotation angles from 0° to 360° are obtained, when observing the C(1')–N bond, for clockwise rotation of the N(9)–C(8) bond of the purines (or the N(1)–C(6) bond of the pyrimidines) relative to the C(1')–O(1') bond [57].

79

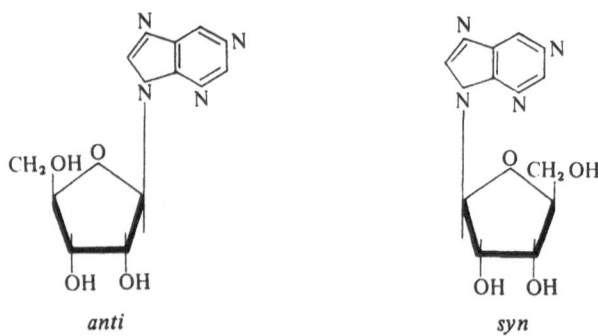

Fig. 23. The *anti* and *syn* conformations of purine nucleoside

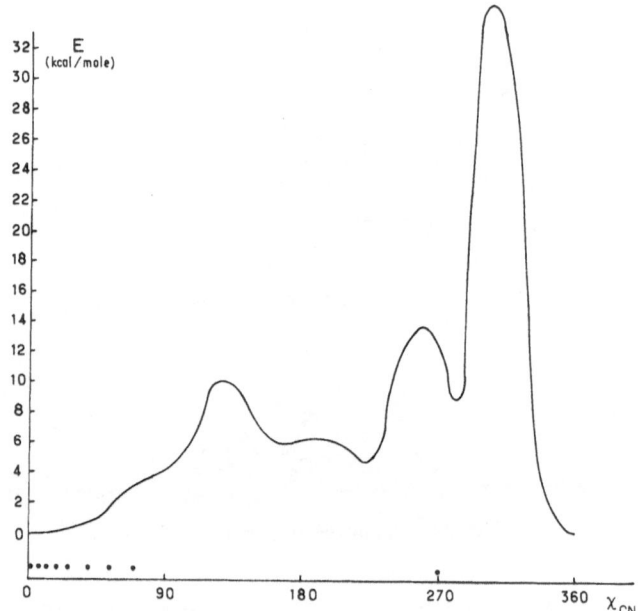

Fig. 24. Conformational map for rotation about the glycosidic bond in C(3')-endo nucleosides of uridine and thymidine

The so-called *"anti"* or *"syn"* conformations of the bases relative to the ribose are determined by the following ranges of χ_{CN}: *"anti"* corresponds to χ_{CN} varying form $0°$ to $\pm 90°$; *"syn"* corresponds to χ_{CN} varying from $180°$ to $\pm 90°$. Pictorially the two types of conformation correspond broadly speaking to the arrangements illustrated in Fig. 23.

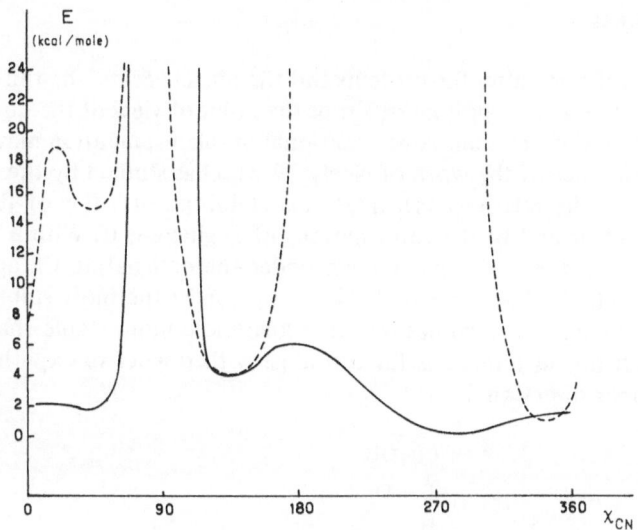

Fig. 25. Conformational energy map for rotation about the glycosidic bond of:
- - - - C(2')-exo-β-adenosine
———— C(2')-exo-α-adenosine

The detailed results for each group of nucleosides depend on the type of puckering of the sugar and the nature (purine or pyrimidine) of the base. The general agreement between theory (PCILO) and experiment is remarkable and indicates an overall preference for the *anti* conformation. This is illustrated in Fig. 24 which represents the conformational energy maps for the χ_{CN} rotations in C(3')-endo nucleosides of uridine and thymidine, the experimental conformation coming from X-ray studies on the corresponding crystals. (For details see [58].)

The study has also thrown light on a number of interesting pecularities of the structure of nucleosides such as for instance, the selective association of the *exo* or *endo* puckerings of the sugars with the α or β anomers. Thus, it was pointed out interestingly by Sundaralingam [59] that the *exo* puckering of the sugar is preferentially associated with the α-nucleosides of purines and pyrimidines. In order to throw some light on the possible reasons for this preference, a calculation has been carried out [60] on the two isomeric C(2')-exo-α-adenosine and C(2')-exo-β-adenosine. The results are indicated in Fig. 25. They indicate that, although the depth of the minima attained is similar for the two anomers, these minima are much narrower for the β-anomer and are separated by very high potential barriers. Altogether the presence of the C(2')-exo-sugar in the α-nucleoside seems thus more probable than its association with a β-nucleoside.

C. Polysaccharides

The polysaccharides are, after the proteins and the nucleic acids, the third important group of biological polymers. From the point of view of the quantum-mechanical calculations of their conformational problems, mention may be made in the first place of the work of Neely [13] who has studied by the Extended Hückel Theory the relative conformational stabilities of a few of the different possible chair and boat conformations of D-glucose, II. Within the conformations studied, the chair one, known under the designation Cl, appears so far the most stable. Experimentally, it is apparently the most stable of all. Moreover, the β-anomer comes out as 9 kcal/mole more stable than the α-one. Although this difference is far too large with respect to experiment, it is in the proper direction.

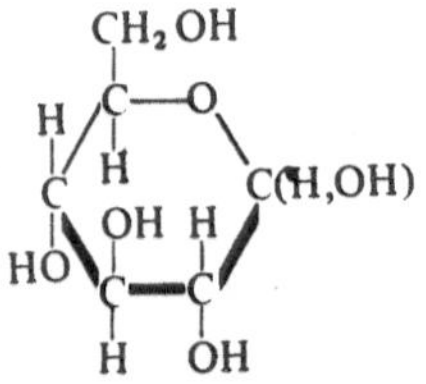

II (α or β)-D-glucose

Much more extensive studies, carried out by the PCILO method, have been centered on the establishment of conformational maps for fundamental dissacharides with a view of their extension to oligo- and poly-saccharides. Two dihedral angles Φ and Ψ are defined around the bridge oxygen atom (Fig. 26), like those used in dipeptides. The available results are significant. Some of their outstanding features may be illustrated by the example of the disaccharide maltose which consists of two glucose residues in the Cl form joined through an α, 1–4' linkage. The conformational map obtained by PCILO calculations is shown in Fig. 27. Its most striking aspect is the very limited area of stable conformations and a rather precise location of the preferred one. (For more details, see [62].)

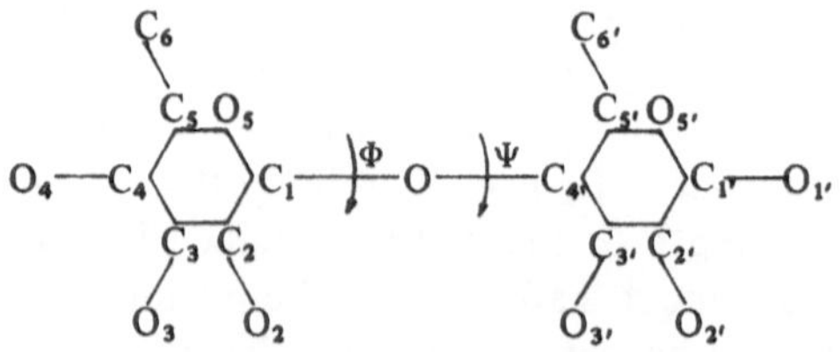

Fig. 26. A dissacharide: two α-β-D-glucose units linked through the bridge oxygen atom. The directions of the rotations Φ and Ψ are indicated

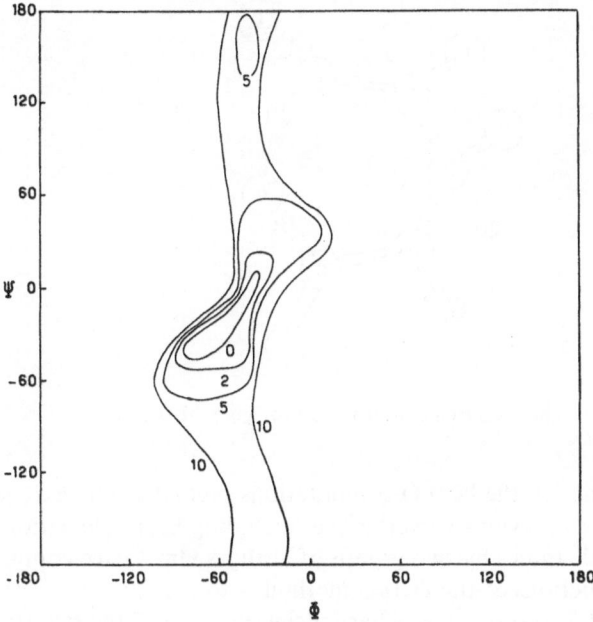

Fig. 27. Conformational map for maltose, in kcal/mole with respect to the global minimum taken as energy zero

D. General Comments

In the preceding examples on the application of quantum-mechanical methods to the study of conformational problems in biochemistry, we have centered our attention essentially on results obtained by the PCILO method. The reason for this situation resides in the first place in the fact that the results obtained by this procedure are by far the most abundant. A second reason is, however, that they appear also the most satisfactory, being in particular superior to those obtained by the Extended Hückel Theory, which comes next after PCILO in the amount of work carried out. (For practical reasons, very few calculations have been performed in this field using the CNDO/2 method.)

A few striking examples may perhaps illustrate this statement.

1) One of the immediate successes of the PCILO method when applied to the simple model dipeptides, say, glycyl-L-alanine, was to predict the *most stable form* of such a compound, a conformation which completely escaped the attention of the great majority of the previous empirical computations and also of the EHT calculations. This conformation is represented by a seven-membered hydrogen-bonded ring (Fig. 28). As the next most stable conforma-

83

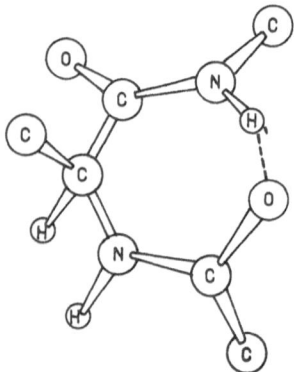

Fig. 28. The preferred conformation for glycyl-L-alanine

tion for the same model, the PCILO computations predict a fully extended form which again had been previously overlooked, including EHT calculations. Now recent experimental studies by two groups of authors [63-66] have completely confirmed the predictions of the PCILO method.

2) The prolyl (PRO) residue is rather special. Because of the rigidity of the pyrrolidine ring, there are strong restrictions about the possible values of the angle Φ, which is limited to $110°-120°$, thus diminishing the allowed space on the conformational map. PCILO calculations (in agreement with the empirical one) indicate two allowed zones for the angle ψ: $130°-170°$ and $300°-350°$ [67]. EHT calculations [68] only allow the second of these zones. In doing so, these last calculations are completely at variance with experimental indications which however, agree with the indications of the PCILO calculations.

3) The conformational energy map of the disulphide bridge of proteins was recently investigated theoretically on the model compound $CH_3-S-S-CH_3$ by Ponnuswamy [53], both by the "empirical" computations and with the help of a simple quantum-mechanical procedure, namely the Extended Hückel Theory. The results of both types of computation were unsatisfactory. Thus, while experimentally the $S-S$ bridges undergo a twist of about $90°$ around the $S-S$ bond in amino acids and in proteins, the "empirical" computations indicate an energy minimum for a rotation of about $35°$ and the Extended Hückel computations favor the angle of $180°$ (*trans* conformation).

PCILO calculations [69] lead to the conformational energy curve of Fig. 29, in complete agreement with experiment.

These few examples do not leave any doubt about the superiority in conformational problems of the PCILO computations over the EHT ones, which even appear sometimes in drastic contradiction with experimental facts. The same situation will be evident in similar studies on pharmacological compounds (Sect. IV).

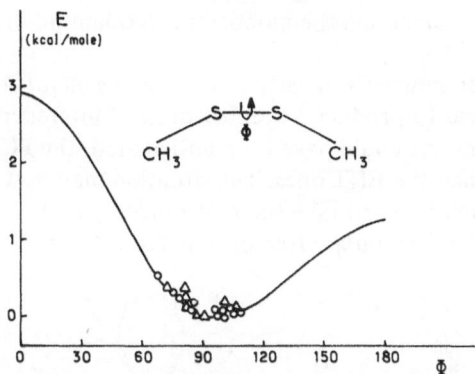

Fig. 29. The conformational energy of $CH_3-S-S-CH_3$ with respect to the energy of the most stable conformation taken as zero. The drawing in the upper rigth corner corresponds to $\Phi = 0°$ (*cis* conformation). Experimental conformations of the $S-S$ bridge in: △ small cystine compounds, □ globular proteins

E. Applications in Pharmacology

The need for a rational approach to pharmacology has become very strong during these last years. Because of the chemical nature of the majority of drugs, which most frequently involve conjugated and saturated fragments and thus imply probably in their mechanism of action the intervention of their σ and π electrons, these compounds have for a long time evaded successful quantum-mechanical treatment. The advent of the all-valence electrons method opens up new vistas in this field.

The pioneering work was carried out by Kier [70,77], who, with the help of the Extended Hückel Theory, has performed calculations on the preferred conformations and the charge distributions in a series of fundamental drugs with the view, primarily, of establishing their pharmacophoric patterns. The drugs studied involved, in particular: acetylcholine, muscarine, nicotine, histamine, ephedrine, norepinephrine, dopamine etc.. This work was summed up recently in a book [78]. A few isolated attempts have also been made by other workers in this field, generally again with the help of the Extended Hückel Theory but sometimes using the CNDO/2 or even the INDO method [79-83].

Recently a systematic investigation of the conformational and electronic aspects of pharmacology has been launched in our laboratory, using the PCILO method. The compounds investigated for which results are already available are those which have been previously investigated by the EHT method: acetylcholine, nicotine and muscarine [84], serotonine [85], histamine [86], ephedrine, norephedrine and dopamine [87], but also others: tyramine, noradrenaline, ephedrine, amphetamine and privine [87], a number of barbiturates [88], and a number

85

of hydrazine derivatives, including the monoamine oxidase inhibitor iproniazid [89] etc.

Although it seems somewhat too early to draw any *general* conclusions from these studies, they appear to produce a number of most interesting results, correlations and predictions. As could have been anticipated, the PCILO results are more satisfactory than the EHT ones. The situation may best be illustrated by an example, for which we shall take the much-investigated case of acetylcholine, III, the natural intercellular effector in nervous transmission systems.

III   Acetylcholine

Although this molecule contains eight single bonds, there are only four important torsion angles (indicated as $\tau_1-\tau_4$ in III), only two of which are essential, i.e. τ_1 and τ_2 [a]. Thus, in the first place the methyl groups of the quaternary nitrogen may be taken in the staggered positions, the trimethyl-ethyl-ammonium backbone being antiplanar with $\tau_3 = \tau(C_5 - C_4 - N^+ - C_3) = 180°$. From our studies on dipeptides we may also fix the C_7-methyl group with a C–H bond eclipsing the $C_6 = O_2$ double bond. Moreover, because of the partial double bond character of the $C_6 - O_1$ bond (similar to the partial double bond character of the C–N bond of the peptide group

$$\overset{\displaystyle O}{\underset{\displaystyle H}{\overset{\displaystyle \|}{C}-N)}}$$

the torsion angle $\tau_0 = \tau(C_7 - C_6 - O_1 - C_5) = 180°$ *(vide infra)*. We are

left therefore with the two essential torsion angles: $\tau_1 = \tau(C_6 - O_1 - C_5 - C_4)$ and $\tau_2 = \tau(O_1 - C_5 - C_4 - N^+)$. The study of these rotations was the center of our attention.

---

[a] Following the usual convention in this type of study in pharmacology, the torsion angle τ of the bonded atoms $A - B - C - D$ is the angle between the planes ABC and BCD. Viewed from the direction of A, τ is positive for clockwise and negative for anticlockwise rotations. The value $\tau = 0°$ corresponds to the planar-*cis* arrangement of the bonds AB and CD. Values of $\tau = 0°, 60°, 120°$ and $180°$ are termed *syn*-planar, *syn*-clinal, *anti*-clinal and *anti*-planar, respectively.

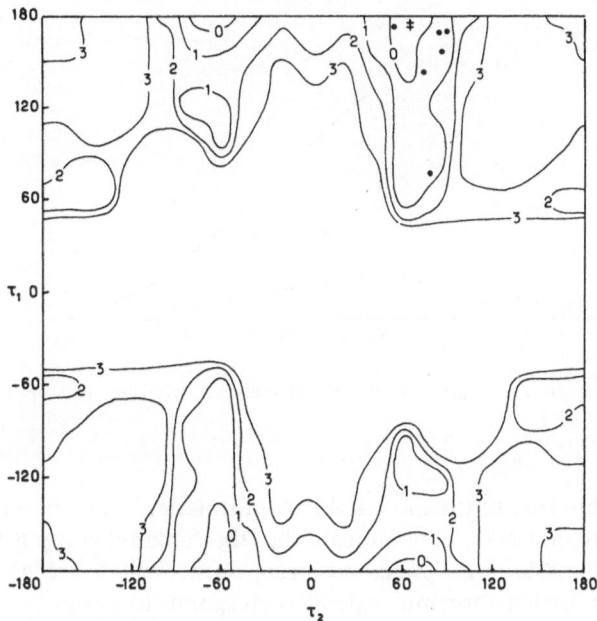

Fig. 30. Conformational energy map of acetylcholine. Isoenergy curves in kcal/mole with respect to the global minimum ‡ taken as energy zero. Experimental conformations ■ in crystals of acetylcholine derivatives

The conformational energy map of acetylcholine (*isoenergy* curves as a function of τ_1 and τ_2) constructed by the PCILO calculations is presented in Fig. 30. There is a global minimum situated at $\tau_1 = 180°$, $\tau_2 = 60°$ which, together with the adopted values of $\tau_0 = \tau_3 = 180°$, corresponds to a structure which may conventionally be described by the symbol TTGT, underlying the *gauche* conformation of the O_1 and N^+ atoms and the *trans* arrangement of the remaining atoms of the backbone. A local minimum, situated at about 1 kcal/mole above the global one, is found at $\tau_1 = 120°$, $\tau_2 = -80°$. (By symmetry, there are of course identical minima at $(\tau_1, \tau_2) = (-180°, -60°)$ and $(-120°, 80°)$ which correspond to enanthiomorphs). It may be usefully stressed that around this minima and in particular around the global one, there are large plateaus of low energy, as illustrated by the relatively broad contour of the 1 kcal/mole isoenergy curve. The minima, and in particular the global one, are therefore associated with a considerable flexibility especially as concerns the variation of τ_1. It may be observed that the experimental conformations of a number of crystals of acetylcholine derivatives agree very satisfactorily with the indications of the calculations.

As concerns the results of the Extended Hückel theory [70], they lead to a preferred conformation associated with $\tau_1 = 180°$ and $\tau_2 = 80°$ and thus close,

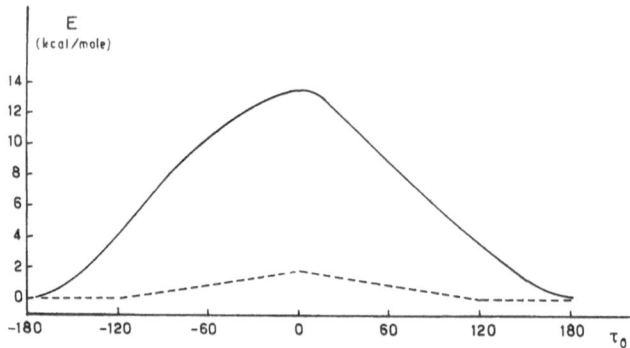

Fig. 31. Energy variation for rotation around τ_0 of acetylcholine ($\tau_1 = 180°, \tau_2 = 60°$)
——— PCILO, ------ EHT

from that point of view, to the most stable conformation found in our calculations. Surprisingly, however, they indicate that the carbonyl group is *free* to rotate 60° to either side of the *planar-cis* arrangement with respect to the $O_1 - C_5$ bond, i.e. that the torsion angle τ_0 corresponds to a constant value of the energy between 120° and 240°. These results are disproved by explicit calculations by the PCILO method (Fig. 31) which show the existence of a definite minimum at 180° corresponding to the planarity of the ester group, the energy rising continuously upon departure from this arrangement. A 60° torsion would correspond to about 4 kcal/mole increase in energy. This divergence between the results of the two procedures has an important significance for the theory of action of this compound, in particular with respect to the origin of the flexibility (if any) of this molecule (for details see Ref. [84]).

IV. The Supermolecule Approach to Hydrogen Bonding

The study of isolated biochemical compounds, however fascinating it may be, reveals only one facet of their reality, the other involving their mutual *interactions*. Among various ways in which these interactions take place, hydrogen bonding is of outstanding importance (nucleic acid base pairing, codon-anticodon recognition, secondary structure of proteins, etc.) and for the theoretician it is perhaps the most fascinating, since it bridges the gap between the familiar domain of chemical bonding and the much less clear area of "long range" forces. For many years, and until very recently, the quantum-mechanical treatment of hydrogen-bonded molecules had been performed either on very truncated systems or by very approximate methods, or both. In the very early stages of the theory, the emphasis had been on the importance of electrostatic forces, but it was soon recognized that short-range repulsions played an important role; then charge transfer and polarization effects where brought into play. However, on account of the approximate or truncated nature of the theoretical treatments utilized, the respective roles and weights of the different contributions to hydrogen bonding remained disputable, an unsatisfactory situation when classification or comparison of different hydrogen bonds is necessary (if one wants to predict, for instance, the relative stabilities of different associations, like those which occur between the base components of the nucleic acids, or between different parts of protein molecules).

Clearly, the most satisfactory way to decide between conflicting concepts of the structure and nature of the hydrogen bond is to treat quantum-mechanically a hydrogen-bonded complex as a single large molecule entity with no truncation and to compare the results obtained for this "supermolecule" to those obtained for the separated molecules treated in the same approximation. This mode of approach is now possible, and a number of such computations using both all-valence electrons methods and the SCF MO non-empirical procedure have recently appeared. The references pertinent to biochemistry have been listed in Tables I and II. These concern only various hydrogen-bonded amides and the base pairs of the nucleic acids.

Owing to their importance as models for hydrogen-bonded peptides in proteins, the formamide dimers have been among the first hydrogen-bonded systems to be dealt with in the supermolecule approach using first the CNDO/2 procedure [90]. A few variants and elaborations of this investigation have been performed since then, generally accompanied by corresponding EHT computations [91,92]. On the other hand, a thorough study of the same dimers has been carried out by the SCF LCAO procedure [93,94] in a completely non-empirical fashion, emphasis being put on the delineation of the respective importance of the different contributions to the binding energy, and to the connection of the supermolecule approach with the classical perturbation treatment of intermolecular forces.

We shall summarize these amide studies to illustrate the new approach.

A. All-Valence Electrons Computations on Hydrogen-Bonded Amides

At the outset of this type of calculation, the first question to be answered was whether the supermolecule method is liable to give reasonable H-bond energies, if any. The first CNDO/2 calculation yielded the values 5.8, 4.9 and 4.6 kcal/mole per hydrogen bond for the dimers IV, V and VI respectively, which was of a satisfactory order of magnitude.

These calculations, however, were done for the *experimental* H-bond distances (that of the crystal structure of IV, and the mean value of those occuring in glycylglycine for V and VI which correspond respectively to the arrange-

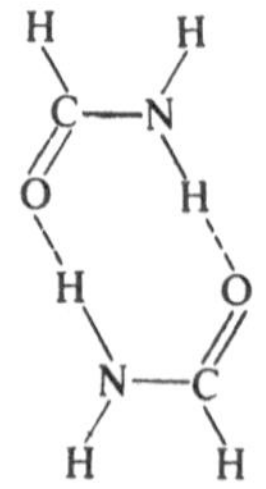

IV

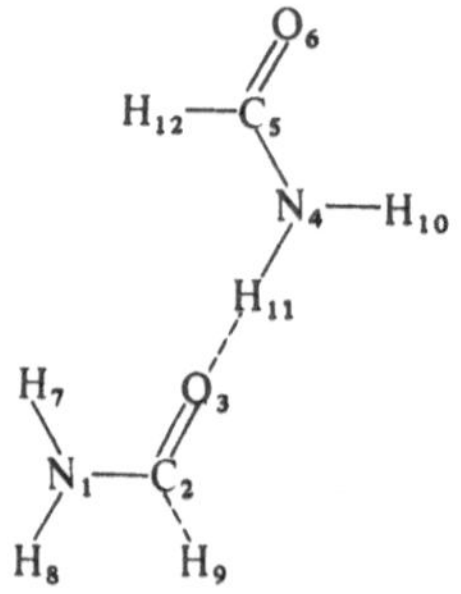

V

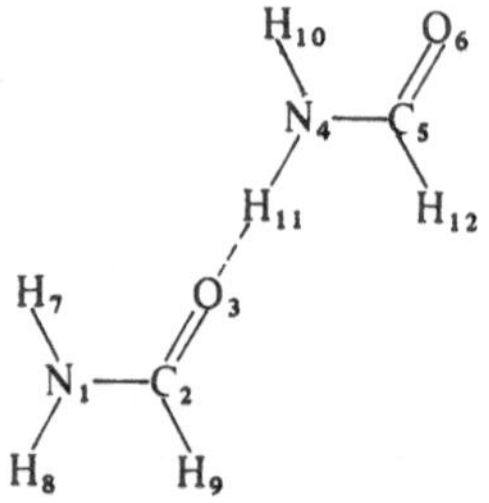

VI

ments occuring in the secondary structure of two antiparallel and parallel poly-peptide chains). When a minimization is performed in terms of the distance of approach of the two monomers, the equilibrium distance is found to be too short by about 0.3 Å and the corresponding H-bond energy too large [91]. This has been recognized as a characteristic feature of the CNDO/2 method, whereas the extended Hückel theory gave no minimum of the energy curve along the path of approach [91-93]. It may be added that the PCILO method improves on the CNDO/2 results both for the position and for the values of the binding energy [94].

The next important question which may be raised concerns the nature of the electron displacements accompanying hydrogen bond formation. This has been discussed abundantly in the past, particularly in connection with the possible role of charge transfer in hydrogen bonding, but the incompleteness of the models utilized did not permit definitive conclusions.

Charge displacements have been studied both by EHT and CNDO. In fact, it has been pointed out [90] that the EHT method is inappropriate for such a study in complexes which involve π as well as σ electrons: the reason for this inadequacy is that the method ignores the effect of the σ displacements on the

π electrons. The very choice of the values of the diagonal elements for the π electrons makes them always the same for a given atom, and thus, the same in the dimer as in the monomer; the equations for the π orbitals for the dimer and monomer thus differ only by the off-diagonal elements connecting couples of atoms in two different units, elements absent in an isolated unit. Their overlap dependence makes these elements so small for π electrons that the π charges remain unaffected.

CNDO/2 presents no comparable inherent defect. The main conclusions referring to the electronic displacements upon hydrogen bond formation emerged immediately in a very clearcut fashion and have not been changed by more recent calculations[a]. They are as follows [90]:

1) Both the nitrogen involved in the hydrogen bond (the proton-donor) and the oxygen on the other side (the proton acceptor) show a global *gain* of electrons. But the nitrogen's overall gain is made up of a σ-gain and a smaller π-loss, while the oxygen presents a small σ-loss and a π-gain.

2) The hydrogen of the hydrogen bond shows a *loss* of electrons.

3) A (very small) transfer of charge occurs *from the proton-accepting unit to the proton-donating unit* (in spite of the increase in the positive charge of the hydrogen atom of this last unit). The bulk of this transfer is essentially σ, changes in π charges appearing thus merely as *rearrangements*.

4) The dipole moments of the linear dimers II and III were found to be larger than the values obtained by simple vector addition of the monomer moments. Similar conclusions were obtained for dimers of N-methyl acetamide [91].

Two other aspects of hydrogen bonding have been studied at the all-valence electron level (in amides): the shape of the proton potential curve and the influence of some angular displacements of one monomer unit around the other. The proton potential curve obtained by CNDO for the formamide and N-methylacetamide dimers has a reasonable shape with a double minimum but the barrier height is too large [91]. The results concerning angular variations are less clearcut and vary somewhat according to method and authors.

B. Non-Empirical Studies on the Dimers of Formamide [95-97]

The possibility of performing non-empirical calculations provides a means of verifying the conclusions reached by the valence-electrons method and offers an opportunity to go further. It must, however, be kept in mind that the results of SCF calculations depend on the atomic basis set adopted, particularly in the

---

[a] The disagreement obtained by Murthy, Rao and Rao as to the charge displacements on the amino nitrogen is clearly due to an error in their π charges (Such a nitrogen cannot be a π acceptor).

systems considered, the size of which precludes the utilization of a very extended basis. Thus, the energy results, in particular, cannot be expected to reproduce the experimental energies of hydrogen bonding. But it may be hoped that relative energy values are meaningful, like other quantities of interest. In fact, a comparative study of the results for two different basis sets has been performed for the cyclic dimer I [95], with the Gaussian basis set utilized for the computations on the nucleic bases mentioned in Sect. I, and with Clementi's basis for conjugated molecules [98]: The stabilization energies were 19 and 14 kcal/mole respectively for two hydrogen bonds at the experimental distance (in the crystals), thus indicating the tendency − further confirmed since − [99] that the better the basis sets, the smaller the hydrogen bond energy. All the other qualitative conclusions were identical in the two basis sets.

A detailed investigation [96,97] was carried out on the "antiparallel" dimer V for:

a) the linear approach of two monomer units, keeping the O, H, N atoms in a line.

b) the angular variations which occur when one unit is rotated around the other at a fixed distance (keeping a coplanar structure).

c) the stretching of the NH bond.

The particular interest of this study resides in the fact that the *components* of the binding energy were put into evidence at each stage, as well as the accompanying electron displacements. Let us briefly see how it is done:

In the SCF supermolecule approach the hydrogen-bond energy is obtained as the difference between SCF dimer energy and twice the energy of a single isolated monomer, calculated in the same way. This SCF energy can be broken down into different contributions, so as to reveal in particular those which would be obtained in a first-order perturbation treatment: let Ψ_M^0 be the SCF wave function of the *isolated* monomer, and suppose that we use as zeroth-order wave function for the dimer a *simple product:*

$$\Psi^0 = \Psi_{M_1}^0 \, \Psi_{M_2}^0$$

In this case, the first-order perturbation energy upon dimer formation would simply be the Coulomb energy E_C between the unperturbed charge distributions on the monomers M_1 and M_2.

If, in a better approach for overlapping molecules, one takes as zeroth-order wave function an antisymmetrized product of the monomer wave functions:

$$\Psi_A^0 = \mathcal{A} \, [\Psi_{M_1}^0 \, \Psi_{M_2}^0]$$

where $\mathcal{A}$ is the antisymmetrizer operator, the first-order energy would become:

$$E^1 = E_C + E_E$$

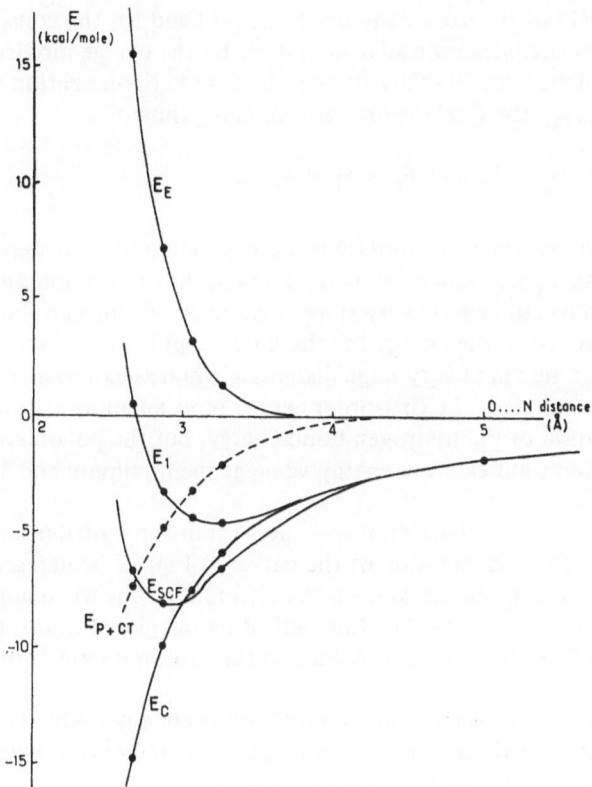

Fig. 32. Energy variation and its components for a linear approach of two formamides

where E_E is the exchange energy due to antisymmetrization. E_C can be computed exactly, knowing the wave functions for M_1 and M_2. Now, E^1 may be obtained as a sub-product of the SCF treatment of the dimer if one uses as trial vectors the SCF molecular orbitals of the isolated monomers [100]. At the first iteration the dimer wave function is indeed simply the antisymmetrized product Ψ_A^0 and E_1 is the corresponding energy. Thus the exchange part of the first-order energy can be obtained by the difference:

$$E_E = E^1 - E_C$$

Finally, the difference between the global SCF energy E_{SCF} and the first-order energy E^1 represents the extra stabilization obtained when pursuing the iteration procedure on the dimer until completion. Thus it accounts on the one hand for polarization and delocalization induced in one unit by the per-

93

manent electric field of the other, and on the other hand for the consequences of modifications in the intramolecular field itself by the charge modifications due to antisymmetrization. Globally this can be denoted polarization + charge-transfer effects E_{P+CT}, the final decomposition being thus:

$$E_{SCF} = E_C + E_E + E_{P+CT}$$

On the other hand, the wave functions corresponding to each step of the energy refinements, E_C, E^1 and E^{SCF} being known, the corresponding electron density curves can be studied at every stage. Fig. 32 shows the variation of the different contributions to the energy for the linear approach: it is seen that the Coulomb attraction sets in at very large distances, whereas exchange repulsions come into play much later. The first-order energy gives an approximate representation of the variation of the hydrogen bond energy, but the polarization and charge transfer effects increase the energy value at the minimum and shorten the equilibrium distance.

The separation of the polarization and charge transfer contributions may be clearly visualized by examination of the curves in Fig. 33, which gives the variations of the electron populations on the different atoms when approaching the two monomers, the broken line indicating the global transfer of population observed from the proton acceptor to the proton donor. It may be observed that:

a) The perturbation of the atomic populations is already visible at large intermolecular distance. This indicates the long-range character of polarization, which sets in early, as does the Coulomb attraction.

b) Transfer of charge starts at a much shorter distance.

c) The variation of σ population of the proton-accepting atom, O_3, shows an inversion of sign when the $O \ldots N$ distance decreases. At large intermolecular separation, the electric field of the NH bond ($N^{\delta-} H^{\delta+}$) polarizes the C=O bond and the oxygen population increases (both σ and π), the more labile π electrons being more affected. At shorter intermolecular separation, this field induces charge transfer from the oxygen lone pair to the other molecule, thus resulting in a loss of σ electrons on O_3.

d) The hydrogen of the hydrogen bond shows a rapid decrease of its electron population.

e) Both atoms at the end of the $O \ldots HN$ bridge show a gain of electron at the equilibrium distance, the gain on N_4 being made of a σ gain and a π loss, while the global gain on O_3 covers a σ loss and a π gain. This is a confirmation of the CNDO results.

f) The transfer of charge is a σ transfer; it is on the whole very small.

Another way of studying the electron displacements due to hydrogen bonding is to draw difference density curves between the dimer and the monomer. Fig. 34 gives such curves in the molecular plane corresponding to two stages

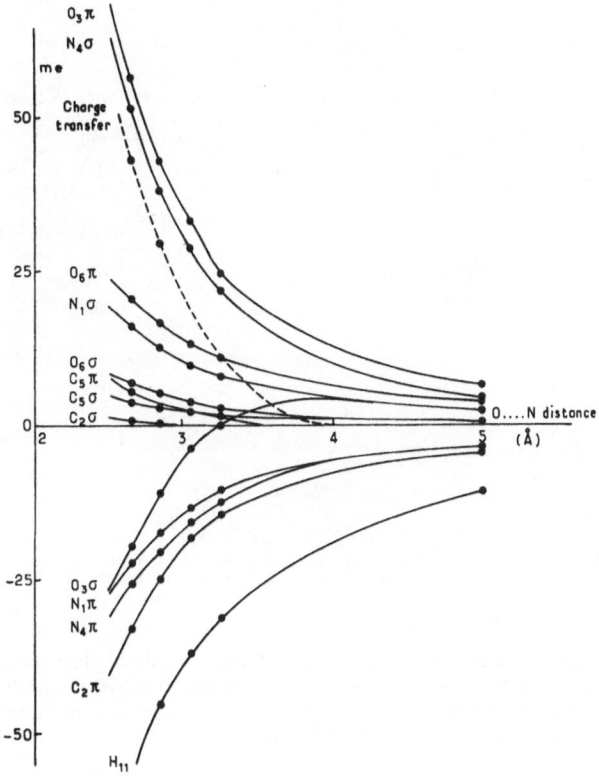

Fig. 33. Gain (> 0) or loss (< 0) in atomic populations and global charge transfer for different distances of two formamide units (10^{-3} atomic units)

in the energy calculation: a) stopping after simple antisymetrization, that is, at the first-order energy level; b) at the SCF stage. It is observed that the effect of antisymetrization is to make the electrons fly away from the intermolecular region (as predicted by Salem [101]). This is a *localized* effect. Then, polarization + charge transfer occurs, bringing about the final modifications, the essential ones being:

a) denuding of the proton,

b) piling-up of charge towards the nitrogen end of the NH bond (increase of the σ-polarity of NH),

c) electron rearrangements *over the whole molecular periphery*,

d) evidence of an overall transfer of sigma density from the proton acceptor to the proton donor.

A similar study has been performed for the in-plane angular rotation of the

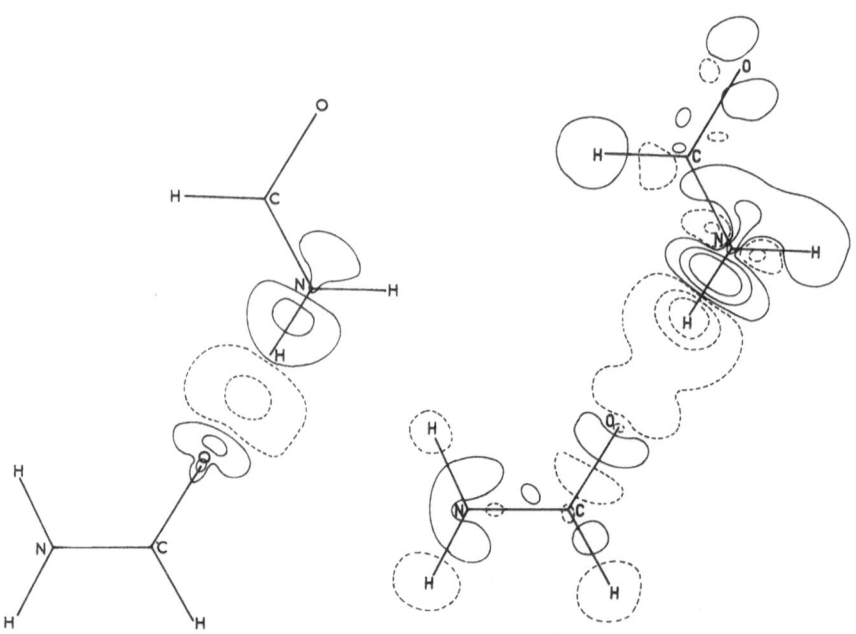

Fig. 34. Isodensity differential maps at equilibrium. Full line: gain of electrons; broken line: loss of electrons. a) Difference between antisymmetrized first-order wave function and two isolated monomers, b) difference between SCF wave function and isolated monomers ($\mp 10^{-3}$ e/a_0^3, outer curve, to $\mp 2.10^{-2}$ e/a_0^3, inner curve)

second monomer around the first, the O....N distances being held fixed at 2.85 Å (Fig. 35).

It is seen that the SCF curve has a very flat minimum between 45° and 75° and a slight shoulder for the linear arrangement. The decomposition of the energy into its components indicates that the electrostatic attraction E_C is quite dissymetrical on each side of $\varphi = 0$, and is the source of the dissymmetry observed in E_{SCF}, the exchange repulsion variation as well as that of E_{P+CT} being symmetrical on each side of $\varphi = 0$. Thus, all contributions to the interaction energy are practically symmetrical on each side of $\varphi = 0$, *apart from the Coulomb energy*. The total energy is minimum for $\varphi \approx 60°$, but a large angular interval is within 1.5 kcal/mole above the minimum. This flatness results from conflicting variations of the different energy components. From a conformational point of view, the antiparallel dimer should adopt a conformation with positive φ, this being entirely due to simple electrostatics. And, indeed, this is observed, for instance, in crystalline N-methyl-acetamide [102].

The flatness of the energy curve is in agreement with the fact outlined by Donohue [22] that the angular conditions required for hydrogen bonding to car-

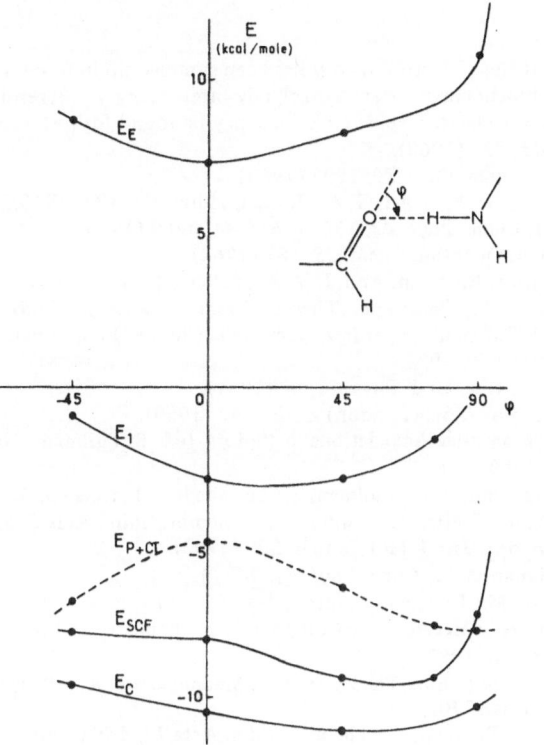

Fig. 35. Energy variation and its components for an in-plane rotation of the second monomer

bonyl groups do not appear to be as stringent as simple considerations on hybridization seem to require. In fact, the lone pair directionality around the oxygen is not as strong as is sometimes thought. (See Sect. II.)

The isodifferential maps have shown that polarization of the proton donor is not very sensitive to φ. The essential features observed in the linear arrangement remain.

The effect of dimerization on the NH-stretching infra-red frequencies has also been studied; the effects calculated are in the right direction (NH lengthening, decrease in the force constant, enhancement of the intensity) the lengthening of the NH bond upon hydrogen bonding appears to be the result of conflicting effects, with exchange forces counteracting the lengthening contributed by Coulomb and polarization effects. Charge transfer alone was shown to be unable to account for the intensity variation [96].

On the whole it appears that more understanding of the phenomenon of hydrogen bonding has been gained by this approach and that further research in this area can be very fruitful.

V. References

[1] The early stage of the π-electron theory has been summed up in B. Pullman and A. Pullman: Quantum Biochemistry. New York:Wiley-Interscience 1963; results of the various refinements in the π-electron approximations may be found for instance in B. Pullman: J. Chem. Phys. *43*, 233 (1965).

[2] Hoffmann, R.: J. Chem. Phys. *39*, 1397 (1963).

[3] Pople, J. A., Santry, D. P., Segal, G. A.: J. Chem. Phys. *43*, 129 (1965).
 – Segal, G. A.: J. Chem. Phys. *43*, 136 (1965); *44*, 3289 (1966).

[4] Pullman, A.: Int. J. Quantum Chem. *2S*, 187 (1968).

[5] – Electron. Aspects Biochem. Ann. N.Y. Acad. Sci. *158*, 65 (1969).

[6] Diner, S., Malrieu, J. P., Claverie, P.: Theoret. Chim. Acta *13*, 1 (1968).

[7] – Appendix to B. Pullman's paper in Aspects de la Chimie Quantique Contemporaine. Paris: Editions du CNRS 1971.

[8] Roothaan, C. C. J.: Rev. Mod. Phys. *23*, 69 (1951).

[9] Boys, S. F.: Proc. Roy. Soc. (London) *A200*, 542 (1950).

[10] Pullman, B., in: Molecular Associations in Biology (ed. B. Pullman). New York: Academic Press 1966.

[11] Lifschitz, C., Bergmann, E. D., Pullman, B.: Tetrahedron Letters *46*, 4583 (1967).

[12] Berthod, H., Giessner-Prettre, C., Pullman, A.: Theoret. Chim. Acta *5*, 53 (1966).

[13] De Voe, H., Tinoco, I.,Jr.: J. Mol. Biol. *4*, 500 (1962).

[14] Berthod, H., Pullman, A.: J. Chim. Phys. *62*, 942 (1965).

[15] Pullman, A., Rossi, M.: Biochem. Biophys. Acta *88*, 211 (1964).

[16] – Quantum Aspects of Heterocyclic Compounds in Chemistry and Biochemistry, II^d Jerusalem Symposium 1970.

[17] – in: Sigma Molecular Orbital Theory (eds. O. Sinanoglu and K. Wiberg). New Haven: Yale University Press 1970.

[18] Giessner-Prettre, C., Pullman, A.: Theoret. Chim. Acta *11*, 159 (1968).

[19] Shillady, D. D., Billingsley, F. P., II, Bloor, J. E.: Theoret. Chim. Acta *21*, 1 (1971).

[20] Pullman, A., Dreyfus, M., Mély, B.: Theoret. Chim. Acta *16*, 85 (1970).

[21] Coppens, P., Sabine, T., Delaplane, R. G., Ibers, J. A.: Acta Cryst. *B25*, 2451 (1969).

[22] Donohue, J.: Structural Chemistry and Molecular Biology (eds. A. Rich and N. Davidson), p. 443. San Francisco: Freeman 1968.

[23] See for instance, Pullman, B., and Pullman, A.: Quantum Biochemistry, loc. cit. ch. III.

[24] Bonaccorsi, R., Petrongolo, G., Scrocco, E., Tomasi, J.: Quantum Aspects of Heterocyclic Compounds in Chemistry and Biochemistry, II^d Jerusalem Symposium 1970.

[25] Bonaccorsi, R., Scrocco, E., Tomasi, J.: J. Chem. Phys. *52*, 5270 (1970).

[26] – – – Theoret. Chim. Acta (Berl.) *20*, 333 (1971).

[27] – – – Theoretical Section Progress Report, Laboratorio di Chimica quantistica del C.N.R. e Istituto die Chimica Fisica dell'Università, p. 35, Pisa 1969–70.

[28] – Pullman, A., Scrocco, E., Tomasi, J.: Theoret. Chim. Acta, *24*, 51 (1972).

[29] Mély, B., Pullman, A.: Theoret. Chim. Acta (Berl.) *13*, 278 (1969).

[30] Nakajima, T., Pullman, A.: J. Chim. Phys. *55*, 793 (1958).

[31] Pullman, A., Pullman, B.: Les Théories Electoniques de la Chimie Organique, for numerous examples. Paris:Masson 1952.

[32] Albert, A., in: Physical Methods in Heterocyclic Chemistry (ed. A. R. Katritzky). New York: Academic Press 1963.

[33] Cavalieri, L. F., Rosenberg, B. H.: J. Am. Chem. Soc. *79*, 5352 (1957). – Zubay, G.: Bioph. Bioch. Acta *28*, 644 (1958).

[34] Christensen, J. J., Rytting, J. H., Izatt, R. M.: Biochem. *9*, 4907 (1970).

[35] Pal, B. C.: Biochem. *1*, 558 (1962).

36) Bryan, R. F., Tomita, K.: Acta Cryst. *15*, 1179 (1962).
37) Brookes, P., Lawley, P. D.: J. Chem. Soc. *1960*, 539; Jones, J. W., Robins, R. K.: J. Am. Chem. Soc. *85*, 193 (1963); Lawley, P. D., Brookes, P.: Biochem. J. *92*, 19C (1964); *89*, 127 (1963).
38) Lawley, P. D.: Progr. Nucl. Acid Res. *5*, 89 (1966).
39) — IVth Jerusalem Symposium, in press.
40) Christensen, J. J., Rytting, J. H., Izatt, R. M.: J. Phys. Chem. *91*, 2700 (1967).
41) Brookes, P., Lawley, P. D.: J. Chem. Soc. (London) *1962*, 1348; Biophys. Biochim. Acta *26*, 450 (1957).
42) Dove, N. F., Wallace, F. A., Davidson, N.: Biochem. Biophys. Res. Commun. *1*, 312 (1959).
43) Giessner-Prettre, C., Pullman, A.: Compt. Rend. Acad. Sci. Paris *272*, C, 750 (1971).
44) Cochran, W.: Acta Cryst. *4*, 81 (1951).
45) Iball, J., Wilson, H. B.: Nature *198*, 1193 (1963).
46) Pullman, B.: Quantum Aspects of Heterocyclic Compounds in Chemistry and Bio-chemistry 4, II^d Jerusalem Symposium 1970.
47) — Pullman, A.: Advances in Heterocyclic Chemistry, in press.
48) — Bergmann, E. D., Weiler-Feilchenfeld, H., Neiman, Z.: Int. J. Quantum Chem. *38*, 103 (1969).
49) Ramachandran, G. N., in: Structural Chemistry and Molecular Biology (ed. A. Rich and N. Davidson), p. 27. San Francisco: Freeman 1968.
50) — Sasisekharen, V.: Advan. Protein Chem. *23*, 283 (1968).
51) Scheraga, H. A.: Advan. Phys. Org. Chem. *6*, 103 (1968).
52) Flory: Statistical Mechanics of Chain Molecules. New York: Wiley-Interscience 1969.
53) Ponnuswamy, P. K.: Thesis. India: University of Madras 1970.
54) Edsall, J. T., Flory, P. J., Kendrew, J. C., Liquori, A. M., Nemethy, G., Ramachandran, G. N., Scheraga, H. A.: Biopolymers *4*, 121 (1966).
55) Maigret, B., Perahia, D., Pullman, B.: J. Theor. Biol. *29*, 275 (1970).
56) Pullman, B., in: Aspects de la Chimie Quantique Contemporaine (eds. R. Daudel and A. Pullman), p. 261. Paris: C.N.R.S. 1971.
57) Sundaralingam, M.: Biopolymers *7*, 821 (1969).
58) Berthod, H., Pullman, B.: Biochem. Biophys. Acta *232*, 595 (1971).
59) Roehrer, D. C., Sundaralingam, M.: J. Am. Chem. Soc. *92*, 4950, 4956 (1970).
60) Berthod, H., Pullman, B.: Biochem. Biophys. Acta *232*, 595 (1971).
61) Neely, W. B.: J. Med. Chem. *12*, 16 (1969).
62) Giacomini, M., Pullman, B., Maigret, B.: Theoret. Chim. Acta *19*, 347 (1970).
63) Bystrov, V. F., Portnova, S. L., Tsetlin, V. I., Ivanov, V. T., Ovchinnikov, Y. A.: Tetrahedron *25*, 493 (1969).
64) Marraud, M., Neel, J., Avignon, M., Huong, P. V.: J. Chim. Phys. *67*, 959 (1970).
65) Avignon, M., Huong, P. V.: Biopolymers *9*, 427 (1970).
66) Maraud, M.: Thesis, University of Nancy 1971.
67) Maigret, B., Perahia, D., Pullman, B.: J. Theoret. Biol. *29*, 275 (1970).
68) Yan, F. J., Momany, F. A., Hoffman, R., Scheraga, H. A.: J. Phys. Chem. *74*, 420 (1970).
69) Perahia, D., Pullman, B.: Biochem. Biophys. Res. Commun. *43*, 65 (1971).
70) Kier, L. B.: Mol. Pharmacol. *3*, 487 (1967).
71) — J. Med. Chem. *11*, 441 (1968).
72) — Mol. Pharmacol. *4*, 70 (1968).
73) — J. Pharmacol. Exp. Therap. *164*, 75 (1968).
74) — J. Pharm. Sci. *57*, 1188 (1968).
75) — J. Med. Chem. *11*, 915 (1968).

[76] Kier, L. B., in: Fundamental Concepts in Drug-Receptor Interactions (eds. Danielli, Moran and Triggle), p. 15. New York: Academic Press 1970.

[77] – in: Aspects de la Chimie Quantique Contemporaine (eds. R. Daudel and A. Pullman), p. 303. Paris: CNRS 1971.

[78] – Molecular Orbital Theory in Drug-Research. New York: Academic Press 1971.

[79] Andrews, P. R.: J. Med. Chem. 12, 761 (1969).

[80] Yonezawa, T., Muro, I., Kato, H., Kimura, M.: Mol. Pharmacol. 5, 446 (1969).

[81] Wohl, A. J.: Mol. Pharmacol. 6, 189, 195 (1970).

[82] Archer, R. A., Boyd, D. B., Demarco, P. V., Tyminski, I. J., Allinger, N. L.: J. Am. Chem. Soc. 92, 5200 (1970).

[83] Green, J. P., Kang, S., in: Molecular Orbital Studies in Chemical Pharmacology (ed. L. B. Kier), p. 105. Berlin-Heidelberg-New York: Springer 1970.

[84] Wohl, A. J., in: Molecular Orbital Studies in Chemical Pharmacology (ed. L. B. Kier), p. 262. Berlin-Heidelberg-New York: Springer 1970.

[85] Pullman, B., Courrière, Ph., Coubeils, T. L.: Mol. Pharmacol. 7, 397 (1971).

[86] Courrière, Ph., Coubeils, J. L., Pullman, B.: Compt. Rend. Acad. Sci. Paris 272, 1697 (1971).

[87] Coubeils, J. L., Courrière, Ph., Pullman, B.: Compt. Rend. Acad. Sci. Paris 272, 1813 (1971).

[88] Pullman, B., Coubeils, J. L., Courrière, Ph., Gervois, J. P.: J. Med. Chem., in press.

[89] – – – J. Theoret. Biol. (in press).

[90] Pullman, A., Berthod, H.: Theoret. Chim. Acta 10, 461 (1968).

[91] Murthy, A. S. N., Rao, K. G., Rao, C. N. R.: J. Am. Chem. Soc. 92, 3544 (1970).

[92] Momany, F. A., McGuir, R. F., Yan, J. F., Scheraga, H. A.: J. Phys. Chem. 74, 2424 (1970).

[93] Murthy, A. S. N., Rao, C. N. R.: J. Mol. Struct. 6, 253 (1970).

[94] Pullman, A., Berthod, H.: In preparation.

[95] Dreyfus, M., Maigret, B., Pullman, A.: Theoret. Chim. Acta 17, 109 (1970).

[96] – Pullman, A.: C. R. Acad. Sc. Paris 271C, 457 (1970).

[97] – – Theoret. Chim. Acta 19, 20 (1970).

[98] André, J. M., André, M. C., Hahn, D., Klint, D., Clementi, E.: Hung. Phys. Acta 27, 493 (1969).

[99] Hankins, D., Moskowitz, J. W., Stillinger, F. H.: J. Chem. Phys. 53, 4544 (1970).

[100] This procedure has been suggested by F. B. Van Duijneveldt. Utrecht: Thesis 1969.

[101] Salem, L.: Proc. Roy. Soc. A264, 379 (1961).

[102] Katz, J. L., Post, B.: Acta Cryst. 13, 624 (1960).

References to Table II

Andrews, P. R.: J. Med. Chem. 12, 761 (1969).
Berthod, H., Pullman, B.: Biochim. Biophys. Acta 232, 595 (1971).
Blizzard, A. C., Santry, D. P.: J. Theoret. Biol. 25, 461 (1969).
Boyd, D. B.: Theoret. Chim. Acta 18, 184 (1970).
– Lipscomb, W. N.: J. Theoret. Biol. 25, 403 (1969).
– – J. Theoret. Biol. 25, 403 (1969).
Caillet, J., Pullman, B.: Theoret. Chim. Acta 17, 377 (1970).

Clementi, E., André, J. M., André, M. C. L., Klint, D., Hahn, D.: Acta Phys. Akad. Sci. Hung. 27, 493 (1969).
– Mehl, J., Von Niessen, W.: J. Chem. Phys. 54, 508 (1971).
Collin, R. L.: Ann. N. Y. Acad. Sci. 158, 50 (1969).
Colombetti, G., Petrongolo, C.: Theoret. Chim. Acta 20, 31 (1971).
Dreyfus, M., Maigret, B., Pullman, A.: Theoret. Chim. Acta 17, 109 (1970).
Fujita, H., Imamura, A., Nagata, Ch.: Bull. Chem. Soc. Japan 42, 1967 (1969).
Giacomini, M., Pullman, B.: Theoret. Chim. Acta 9, 347 (1970).
Govil, G.: J. Chem. Soc. (A) 1970, 2464.
– J. Chem. Soc. (A) 1971, 380.
Hoffmann, R., Imamura, A.: Biopolymers 7, 207 (1969).
Imamura, A., Fujita, H., Nagata, Ch.: Bull. Chem. Soc. Japan 42, 3118 (1969).
Jordan, F., Pullman, B.: Theoret. Chim. Acta 9, 242 (1968).
Kang, S., Olsen, T. F., Hamann, J. R.: J. Theoret. Biol. 28, 195 (1970).
Kier, L. B.: J. Med. Chemistry 11, 915 (1968).
– George, J. M., in: Molecular Orbital Studies in Chemical Pharmacology (ed. L. B. Kier), p. 82. Berlin-Heidelberg-New York: Springer 1970.
Langlet, J., Pullman, B., Berthod, H.: J. Chim. Phys. 67, 480 (1970); J. Mol. Struct. 6, 139 (1970).
– – Dreyfus, M.: J. Theoret. Biol. 26, 321 (1970).
– – Perahia, D.: Biopolymers 10, 107, 491 (1971).
Mély, B., Pullman, A.: Theoret. Chim. Acta 13, 278 (1969).
Moffat, J. B.: J. Theoret. Biol. 26, 437 (1970).
Neely, B., in: Molecular Orbital Studies in Chemical Pharmacology (ed. L. B. Kier), p. 121. Berlin-Heidelberg-New York: Springer 1970.
Pullman, A., in: Modern Quantum Chemistry (ed. O. Sinanoglu), Vol. III, p. 283. New York: Academic Press 1965.
– Int. J. Quantum Chem. 2S, 187 (1968).
– Ann. N. Y. Acad. Sci. 158, 65 (1969).
– in: Quantum Aspects of Heterocyclic Compounds in Chemistry and Biochemistry (eds. E. D. Bergmann and B. Pullman), p. 9. Israel Academy of Sciences and Academic Press 1970.
– Berthod, H.: Theoret. Chim. Acta 10, 461 (1968).
Pullman, B., in: Quantum Aspects of Heterocyclic Compounds in Chemistry and Biochemistry (eds. E. D. Bergmann and B. Pullman), p. 292. Israel Academy of Sciences and Academic press 1970.
– in: Aspects de la Chimie Quantique Contemporaine (eds. R. Daudel and A. Pullman), p. 261. Paris: CNRS 1971.
– Pullman A.: Progr. Nucl. Acid Res. and Mol. Biol. 9, 327 (1969).
– Langlet, J., Berthod, H.: J. Theoret. Biol. 23, 492 (1969).
– Maigret, B., Perahia, D.: Theoret. Chim. Acta 18, 44 (1970).
Rein, R., Fukuda, N., Clarke, G. A., Harris, F. E.: J. Theoret. Biol. 21, 88 (1968).
Robb, M. A., Csizmadia, I. G.: J. Chem. Phys. 50, 1819 (1969).
Rossi, A., David, C. W., Schor, R.: Theoret. Chim. Acta 14, 429 (1969).
Snyder, L. C., Schulman, R. G., Neumann, D. B.: J. Chem. Phys. 53, 256 (1970).
Song, P. S.: Ann. N. Y. Acad. Sci. 158, 410 (1969).
Wohl, A. J.: Proceedings of the Seattle Conference on Quantum Biology (Oct. 1969), p. 262. Berlin-Heidelberg-New York: Springer 1970.
Yan, J. F., Momany, F. A., Hoffmann, R., Scheraga, H. A.: J. Phys. Chim. 74, 420 (1970).
Yonezawa, T., Muro, I., Kato, H., Kimura, M.: Mol. Pharmacol. 5, 446 (1969).
Zerner, M., Gouterman, M.: Theoret. Chim. Acta 4, 44 (1966).

A. Pullman

References to Table IX

Berthod, H., Pullman, B.: Biochim. Biophys. Acta 232, 595 (1971).
Bossa, M., Damiani, A., Fidenzi, R., Gigli, S., Leli, R., Ramunni, G.: Theoret. Chim. Acta 20, 299 (1971).
Boyd, D. B., Lipscomb, W. N.: J. Theoret. Biol. 25, 403 (1969).
Caillet, J., Pullman, B.: Theoret. Chim. Acta 17, 383 (1970).
Collin, R.: Ann. N. Y. Acad. Sci. 158, 50 (1969).
Dreyfus, M., Pullman, A.: Theoret. Chim. Acta 19, 20 (1970).
Flory, P. J.: Statistical Mechanics of Chain Molecules. New York: Wiley-Intersciences 1969.
Giacomini, M., Maigret, B., Pullman, B.: Theoret. Chim. Acta 19, 347 (1970).
Goebel, C. V., Dimpft, W. L., Brant, D. A.: Macromolecules 3, 644 (1971).
Govil, G., Saran, A.: J. Theoret. Biol. 30, 621 (1969).
− J. Chem. Soc. (A) 1970. 2464.
− J. Chem. Soc. (A) 1971, 386.
Haschemeyer, A. E. V., Rich, A.: J. Mol. Biol. 27, 369 (1967).
Imamura, A., Fujita, H., Nagata, Ch.: Bull. Chem. Soc. Japan 42, 3118 (1969).
Jordan, F., Pullman, B.: Theoret. Chim. Acta 9, 242 (1968).
Kier, L. B., George, J. M., in: Molecular Orbital Studies in Chemical Pharmacology (ed. L. B. Kier), p. 82. Berlin-Heidelberg-New York: Springer 1970.
Langlet, J., Pullman, B., Berthod, H.: J. Mol. Structure 10, 139 (1970).
− − − J. Chim. Phys. 67, 480 (1970).
Liquori, A.: Quater. Rev. Biophys. 2, 65 (1969).
Maigret, B., Pullman, B., Dreyfus, M.: J. Theoret. Biol. 26, 321 (1970).
− Perahia, D., Pullman, B.: J. Theoret. Biol. 29, 275 (1970).
− Pullman, B., Perahia, D.: Biopolymers 10, 107; 10, 491 (1971).
Momany, F. A., McGuire, R. F., Yan, J. F., Scheraga, H. A.: J. Phys. Chem. 74, 2424 (1970).
Nash, H. A.: J. Theoret. Biol. 22, 314 (1969).
Neely, B., in: Molecular Orbital Studies in Chemical Pharmacology (ed. L. Kier), p. 121. Berlin-Heidelberg-New York: Springer 1970.
Poland, D., Scheraga, H. A.: Biochemistry 6, 3791 (1967).
Ponnuswamy, P. K.: Thesis. India: University of Madras 1970.
Popov, E. M., Doshevskii, V. G., Lipkind, G. M., Arkhipova, S. F.: Mol. Biol. (URSS) 2, 491 (1968).
− Lipkind, G. M., Arkhipova, S. F., Dashevskii, V. G.: Mol. Biol. (URSS) 2, 498 (1968).
Pullman, A., Berthod, H.: Theoret. Chim. Acta 10, 461 (1968).
Pullman, B., Langlet, J., Berthod, H.: J. Theoret. Biol. 23, 492 (1969).
− Maigret, B., Perahia, D.: Theoret. Chim. Acta 18, 44 (1970).
− in: Aspects de la Chimie Quantique Contemporaine (eds. R. Daudel and A. Pullman), p. 261. Paris: C.N.R.S. 1971.
Ramachandran, G. N., Venkatachalam, C. M., Krimm, S.: Biophys. J. 6, 849 (1966).
− in: Structural Chemistry and Molecular Biology (eds. A. Rich and N. Davidson), p. 27. San Francisco: Freeman and Co. 1968.
− Sasisekharan, V.: Adv. Protein Chemistry 23, 283 (1968).
Rao, V. S. R., Sundararajan, P. R., Ramakrishnan, C., Ramachandran, G. N.: Conformation of Biopolymers (ed. Ramachandran), p. 721. New York: Academic Press 1967.
− Yathindra, N., Sundararajan, P. R.: Biopolymers 8, 325 (1969).
Rees, D. A., Skerett, R. J.: Carbohydrate Res. 7, 334 (1968).
− Scott, W. E.: J. Chem. Soc. 13, 469 (1971).
Rossi, A. R., David, C. W., Schor, R.. Theoret. Chim. Acta 14, 429 (1969).

Quantum Biochemistry at the All- or Quasi-All-Electrons Level

Rossi, A. R., David, C. W., Schott, R.: J. Phys. Chem. 74, 4551 (1970).
Sasisekharan, V., Lakshminarayanan, A. V., Ramachandran, G. N.: Conformation of Biopolymers (ed. Ramachandran), Vol. 2, p. 641. New York: Academic Press 1967.
― ― Biopolymers 8, 475, 489, 505 (1969).
Scheraga, H. A.: Adv. Phys. Org. Chem. 6, 103 (1968).
Sundararajan, P. R., Rao, V. S. R.: Tetrahedron 24, 289 (1968).
― ― Biopolymers 8, 305, 313 (1969).
Tinoco, I., Jr., Davis, R. C., Jaskunas, S. R.: Molecular Associations in Biology (ed. B. Pullman), p. 77. New York: Academic Press 1968.
Yan, J. F., Momany, F. A., Hoffmann, R., Scheraga, H. A.: J. Phys. Chem. 74, 420 (1970).

Received July 7, 1971

The Theory of Radiationless Processes in Polyatomic Molecules*

Professor Dr. Karl F. Freed**

The Department of Chemistry and The James Franck Institute, The University of Chicago, Chicago, Illinois, USA

Contents

* This research is supported in part by National Science Foundation Grant GP-28 135 and a grant from the Xerox Corporation, and has benefitted from the use of facilities provided by the Advanced Research Projects Agency for materials research at the University of Chicago.
** Alfred P. Sloan Foundation Fellow.

1. Introduction

In attempting to understand the properties of, and/or processes involving, excited electronic states in polyatomic molecules, it is often necessary to consider radiationless transitions. Radiationless processes can be classified into a number of types:

   (a) processes which involves the breaking or rearrangement of chemical bonds;
   (b) processes in which there is ionization; and lastly
   (c) processes in which there is neither ionization nor breaking of chemical bonds.

The third kind of radiationless processes, the radiationless transitions, are the subject of this review. The first type, photochemical processes, are a generalization of the simple radiationless transitions and are only briefly discussed. It should become apparent that a thorough understanding of the simpler radiationless transitions is an important prerequisite to the development of a satisfactory quantum mechanical theory of photochemical reactions. The relevance of the study of simple radiationless processes to a description of photochemical reactions does not arise solely from the fact that both processes are often observed simultaneously.

In the review we discuss the recent developments which have led to a unified quantum mechanical theory of all of the diverse phenomena which are classified as involving radiationless processes. Since the recent developments and the resultant description of the radiationless processes are the main interest, reference to the older work is made only insofar as it is necessary for the sake of clarity and completeness. Except for some (very interesting) recent developments, the experimental basis for our understanding of radiationless processes is reviewed by Jortner, Rice, and Hochstrasser [1], so it need not be repeated here, except to illustrate some typical behavior.

A number of controversies have arisen in the literature concerning some aspects of the theory of radiationless processes, and an attempt is made to clarify a number of these matters. In the next section we discuss the simple molecular model which is sufficiently general to enable its use in the description of radiationless transitions.

2. The Model

Fig. 1 presents a molecular energy level diagram which has served as the basis for the understanding of radiationless transitions in polyatomic molecules. The molecule under consideration has a set of ground vibronic states which may be thermally populated under typical experimental conditions. A generic ground vibronic state has the wave function ϕ_0 in the Born-Oppenheimer (BO) approximation. The molecule also has an excited electronic state, one of whose

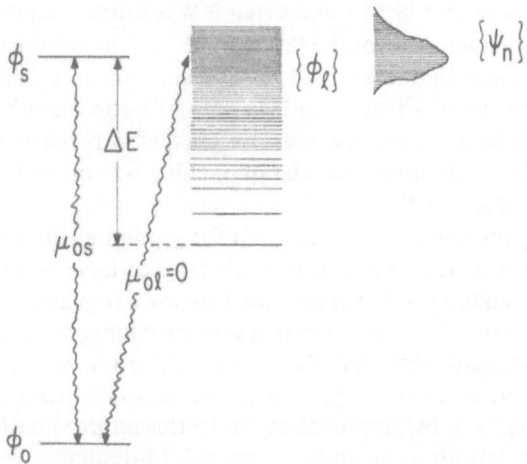

Fig. 1. The molecular energy level model used to discuss radiationless transitions in polyatomic molecules. ϕ_0, ϕ_s, and $\{\phi_l\}$ are vibronic components of the ground, an excited, and
a third electronic state, respectively, in the Born-Oppenheimer approximation. ϕ_s and ϕ_0
are isoenergetic states which are coupled by the terms (effective matrix elements) which
are neglected in the Born-Oppenheimer approximation. Optical transitions between ϕ_0 and
ϕ_s are assumed to be allowed, while transitions between $\{\phi_l\}$ and the thermally accessible
ϕ_0 are assumed to be forbidden. The $\{\psi_n\}$ are the molecular eigenstates

vibronic components ϕ_s (depicted in Fig. 1) carries nonzero oscillator strength
for electric dipole transitions from the ground electronic state ϕ_0. The wave
function ϕ_s is also taken to be that given by the BO approximation. Isoenergetic with ϕ_s is a dense manifold of vibronic states $\{\phi_l\}$ which correspond to highly vibrationally excited levels of some lower electronic state, not excluding the
ground electronic state. The $\{\phi_l\}$ are taken as BO functions and it is assumed that
electric dipole transitions from ϕ_0 to $\{\phi_l\}$ are essentially forbidden either because of spin symmetry rules or because of very unfavorable Franck-Condon
factors [2]. These states $\{\phi_l\}$ may then be able to fluoresce or phosphoresce to
excited vibronic components of the ground electronic state. It is assumed that
photochemical decompostion is not involved.

Luminescence experiments were initially performed upon large molecules
within this general scheme of molecular energy levels for cases in which the
molecules were in solution or were trapped in rigid media. In these cases light
containing frequencies corresponding to the transition energy between ϕ_0 and
ϕ_s is incident upon the system. It is then assumed that the molecule is excited
to state ϕ_s, since that is the only excited molecular state in the region which
carries oscillator strength from ϕ_0. The luminescence from ϕ_s is monitored by
observing the decay of the fluorescence, i.e. one measures the fluorescence
lifetime, τ_f, and/or the quantum yield for fluorescence, φ_f. The basic experi-

mental observation was that $\varphi_f < 1$ and often it was found that $\varphi_f \ll 1$. The process leading to the quenching of the fluorescence was determined to be primarily of an intramolecular character. Thus, anthracene appears to have the same τ_f and φ_f in low pressure gases and a wide variety of hydrocarbon solutions [3]. τ_f for bracetyl is the same in the low pressure gas and in benzene solution [4], etc. The rationale for such intramolecular quenching was taken to be the breakdown of the approximation [5].

While the BO approximation is very good for ground electronic states, it is not an adequate representation of the excited states of large molecules. The terms neglected in making the BO approximation are often small; however, *whenever degenerate or nearly degenerate levels are connected by a perturbation, no matter how small, this perturbation has important qualitative effects.* For the sake of this discussion the spin-orbit interaction is included among those terms which "violate" the BO approximation. In this approximation the BO states are ones with definite spin, and the spin orbit interaction can lead to couplings between these different pure spin BO molecular states. For the present no specification is given as to which "version" of the BO approximation is to be taken (see Sect. 9). The present section considers only a mechanistic description of the radiationless transitions. In both popular versions of the BO approximation, the molecular level scheme corresponds to that presented in Fig. 1, so the discussion applies to both of these cases. The dipole strength μ_{0s} and the interaction energy coupling ϕ_s and $\{\phi_l\}$, v_{sl}, are taken as effective values reflecting vibronic coupling effects due to other BO electronic states which are not isoenergetic with ϕ_s and $\{\phi_l\}$. These effective parameters are discussed more fully in Sect. 9.

The breakdown of the BO approximation leads therefore to interactions which couple ϕ_s with $\{\phi_l\}$ [5-10]. Thus a molecule "initally in ϕ_s," upon excitation by the incident radiation, may either fluoresce or "cross-over" to $\{\phi_l\}$. In rigid media or in solutions, ϕ_s would be a vibrationally relaxed state, while the $\{\phi_l\}$ represent vibrationally "hot" states. The latter are quickly vibrationally relaxed by the surrounding medium. Thus, if ϕ_s represents the lowest excited molecular singlet state S_1 and $\{\phi_l\}$ are excited vibronic components of the lowest triplet state T_1, phosphorescence from the lowest levels of T_1 to the ground state S_0 is often observed in aromatic hydrocarbons. This vibrational relaxation of the $\{\phi_l\}$ to states of considerably lower energy than ϕ_s prevents the molecule from "crossing-back" from $\{\phi_l\}$ to ϕ_s. When ϕ_s and $\{\phi_l\}$ have different spin multiplicities this "crossing-over" is called *intersystem crossing*, while the term internal conversion is employed for cases in which the spin multiplies are the same.

In molecules such as the aromatic hydrocarbons for which the most experimental data are available, it was found that the fluorescence lifetimes τ_f were much shorter than that which could be deduced from the integrated absorption intensity. The latter provides the oscillator strength f, and a radiative lifetime

τ_R is associated with f via $\tau_R \propto 1/f$. The decrease in τ_f relative to τ_R was explained in terms of the radiationless transition from ϕ_s to $\{\phi_l\}$. Thus, ϕ_s decays by two processes — radiatively with rate τ_R^{-1} and non-radiatively with rate $\tau_{n.r.}^{-1}$. Thus

$$\varphi_f \equiv \frac{\tau_R^{-1}}{\tau_R^{-1} + \tau_{n.r.}^{-1}} = \frac{\tau_{n.r.}}{\tau_{n.r.} + \tau_R}, \tag{1}$$

and

$$\tau_f^{-1} \equiv \tau_R^{-1} + \tau_{n.r.}^{-1}. \tag{2}$$

describe the observed fluorescence quantum yields and lifetimes in terms of the radiative and nonradiative lifetimes. Similar conclusions were obtained for phosphorescence quantum yields φ_p and lifetimes τ_p in terms of the radiative and nonradiative lifetimes, τ_R^p and $\tau_{n.r.}^p$.

Although the radiationless processes appeared to be intramolecular in character, the presence of the solvent or rigid matrix presented an essential complication to a theoretical description of the radiationless phenomena. A complete theory of radiationless processes must begin with a description of electronic relaxation in *isolated* molecules. Once the isolated molecule case is properly understood, the effects of the external media can be considered and the bulk of the experimental data concerning radiationless transitions can be confronted.

The results that were expected in isolated molecules can easily be summarized as follows: A molecule "initially excited" to ϕ_s may fluoresce with some probability p_f, or "cross-over" to $\{\phi_l\}$ with probabilty $1 - p_f$. Because the coupling v_{sl} between ϕ_s and $\{\phi_l\}$, the levels $\{\phi_s, \phi_l\}$ represents a finite set of molecular states which are "in resonance," using the traditional chemical sense of the term resonance. Thus, when the molecule is in $\{\phi_l\}$, it will eventually "cross-back" to ϕ_s. (Energy must be conserved in all "crossings" in truly isolated molecules.) Once in ϕ_s the molecule may again fluoresce or "cross-over" with probabilities p_f and $1 - p_f$, respectively, etc. Continued to the limit, this argument implies that the total fluorescence probability, the fluorescence quantum yield, must be unity in an isolated molecule. An identical argument holds in the case of phosphorescence.

Kistiakowsky and Parmenter considered the fluorescence quantum yield of benzene at pressures (0.01 torr) which were low enough such that no pressure dependence was observed [11]. Subsequent work has gone to lower pressures with no observed pressure dependence [12]. Indeed, the low pressure result they obtained was that φ_f was not unity, but was $\cong .34$ opposed to the value $.18$ which was found at high pressured and in "inert" solutions. Kistiakowsky and Parmenter remarked that their result seemed to contradict the laws of quantum mechanics [11].

We note that in the case of benzene ϕ_s corresponds to S_1 while the $\{\phi_l\}$ are vibrationally excited levels of T_1. Arguments were then presented which noted that the true molecular eigenstates $\{\psi_n\}$ were not pure singlet or triplet states in benzene but were admixtures of singlet and triplet. Thus, a molecule "initially excited" to ϕ_s would initially be in a nonstationary state of the full molecular Hamiltonian H_M. This nonstationary state ϕ_s could be represented as a superposition of stationary molecular eigenstates which would evolve in time according to the full (exact) molecular Hamiltonian, H_M [10,13-15]. The probability that the molecule would be in ϕ_s (the initial nonstationary state) at a later time was then found to decay exponentially for short times with lifetime $\tau_{n.r.}$. The molecular eigenstates are depicted schematically in Fig. 1. Since each molecular eigenstate has a component of ϕ_s, it carries part of the total oscillator strength f to the extent that it contains ϕ_s.

The arguments presented above for the BO states can also be extended to the molecular eigenstates to again imply that the fluorescence (or phosphorescence) probability in an isolated molecule must be unity [16]. As noted in Sect. 7, although the above discussion of radiationless decay in terms of molecular eigenstates had considerable pedagogical value, *when we admit of the possibility of spontaneous emission of radiation (fluorescence and phosphorescence), the molecular eigenstates lose their physical simplicity* (they no longer "diagonalize" the problem) [14-18] for these large polyatomic molecules.

To compound the confusion, Douglas noted that in small polyatomic molecules such as NO_2 and SO_2, the observed values of τ_f were very much larger than the radiative lifetimes τ_R that would be deduced from the integrated absorption intensity or f [19]. It is well known that the absorption spectra of these molecules contain a large number of unexpected additional lines, and to date they have resisted satisfactory assignment. The molecular level scheme is expected to correspond to the model presented in Fig. 1, and Douglas explained the extra lines and long lifetimes τ_f as arising from the coupling energy v_{ls} — the berakdown of the BO approximation. If a number of levels share the original oscillator strength f, they each have a fraction of f and hence a longer lifetime than the τ_R which is obtained from the integrated absorption intensity. The experimental data for these small polyatomic molecules are also not inconsistent with unit fluorescence quantum yields [20].

In summary, the breakdown of the BO approximation in isolated large polyatomic molecules experimentally leads to

$$\varphi_f < 1, \ \tau_f < \tau_R \qquad (3)$$

while in small molecules

$$\varphi_f \approx 1, \ \tau_f > \tau_R \qquad (4)$$

is observed. (See Sect. 4 for a discussion of irreversibility.)

3. Goals of the Theory

In formulating a theory of radiationless processes in polyatomic molecules, it is important to separate the development into discussions of general principle and the treatment of specific details. In the former category we wish to explain all of the diverse phenomena, which are classified as involving radiationless phenomena in polyatomic molecules, within the framework of a single unified quantum mechanical theory [16,17]. Thus, it is necessary to explain how a dense manifold of discrete states $\{\phi_l\}$ can lead to irreversible relaxation in isolated polyatomic molecules. Once the irreversibility paradox has been answered, the general criteria for irreversible behavior, i.e., irreversible electronic relaxation, in polyatomic molecules should be examined. In particular, it is necessary to determine how these criteria depend upon the coupling strengths v_{sl} between ϕ_s and $\{\phi_l\}$ and upon the density of states $\rho_l(E_s)$ of the $\{\phi_l\}$ in the energy region of ϕ_s. The criteria for nonradiative decay should explicity exhibit any dependence upon whether the radiationless decay involves intersystem crossing or internal conversion. It is also imperative that the theory explain the anomously long lifetimes of small molecules. Finally, once a general understanding of radiationless transitions is obtained for isolated molecules, it is important to extend this theory to include the effects of the external medium, in order to confront the vast majority of the experimental data.

From theoretical discussions involving the molecular eigenstates picture questions have arisen as to whether particular quantum mechanical interference effects can be observed by the use of suitably monochromatic radiation for excitation of the molecules [13]. (See Sect. 7.) Of course, it is also necessary to settle the controversies as to whether the BO or molecular eigenstates are correct, and if the former is indeed correct, which particular version of the BO approximation is to be employed for the calculation of nonradiative decay rates.

There is a great deal of experimental data available concerning the nonradiative decay processes in aromatic hydrocarbons. Siebrand noted that a plot of the logarithm of the intersystem crossing rate $k_{n.r.}$ for $T_1 \rightarrow S_0$ transitions in aromatic hydrocarbons versus ΔE, the separation between the minima of these two potential surfaces, leads approximately to a linear realtionship [6,9,21,22]. His correlation is presented in Fig. 2. It should be noted that the aromatic hydrocarbons represent a homologous series, so there is no *a priori* reason to expect such a good correlation. In Fig. 2 the $k_{n.r.}$ for the perdeutero compounds are also presented. Again the correlation is apparent with a marked isotope effect; the deuterated rates are smaller and have a slope of greater magnitude. It is imperative that the theory provide a quantitative understanding of these trends. Recent experiments have investigated the dependence of $k_{n.r.}$ upon the initially selected vibronic state in benzene S_1 [23-26]. Such experimental data are expected to provide a still more stringent test of the quantitative theory.

111

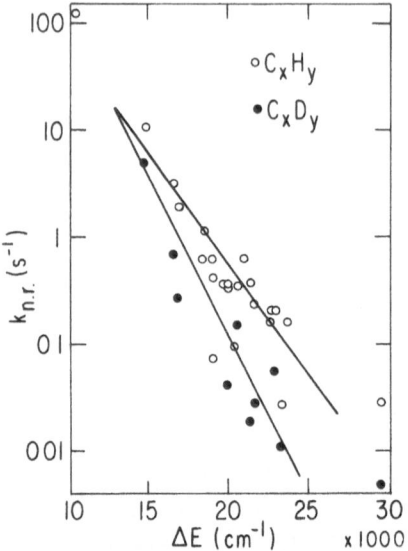

Fig. 2. The dependence of the nonradiative decay rate for intersystem crossing (from the $T_1 \rightarrow S_0$) on the electronic energy gap for perhydro and perdeutero aromatic hydrocarbons. All data were obtained in solid solutions. The nonradiative decay rate was calculated by Siebrand [9b] from Eqs. (1) and (2) taking $\tau_R = 30$ sec. This figure is reproduced from Siebrand's paper [9b], with the added straight lines

The theory should also be capable of discussing such effects of dense media as the temperature dependence of $k_{n.r.}$ and absorption linewidths. Lastly, it is of great chemical importance to extend the theory in order to describe photochemistry as a radiationless process.

4. Irreversibility [16]

The first question posed by the experimental data on electronic relaxation concerns the rationalization of the apparent irreversible nonradiative decay. In solutions and in rigid media Robinson noted that the presence of the surrounding medium provides a heat bath which assures irreversible electronic relaxation (because of the subsequent degradation of the excess vibrational energy) [6]. He also argued that in an isolated large molecule the density of states $\rho_l(E_s)$ is sufficiently large that "the molecule can act as its own heat bath." [6] The arguments presented in Sect. 2 would imply that an isolated molecule cannot be its own heat bath!

This irreversibility paradox arose slightly less than a hundred years ago in a different context. It will be recalled that Boltzmann used plausible physical assumptions to develop a kinetic equation which described the nonequilibrium behavior of a dilute classical gas [27]. This equation predicts that an initial non-equilibrium distribution of interacting structureless particles will decay to the appropriate equilibrium Maxwell-Boltzmann distribution. There is a recurrence theorem (due to Poincaré), however, which states that if we wait long enough, any finite system obeying the laws of classical mechanics will ultimately return (arbitrarily closely) to its initial state. Now, we know from experience that isolated systems tend to equilibrium, e.g., an isolated system in which all of the particles are in one-half of a macroscopic container, after removal of the partition, will eventually tend towards an equilibrium distribution in which the particles are uniformly distributed throughout the box. The recurrence theorem would imply that if we waited sufficiently long, the system would spontaneously go back to its initial state with the particles in one-half of the box. Boltzmann's equation was therefore criticized because it did contain this recurrence which is a direct consequence of the laws of classical mechanics. Boltzmann's reply was basically: "You wait for the recurrences!" For, any estimate of the time scale involved gives a result which is many orders of magnitude longer than the age of the universe.

In a discussion of irreversibility and radiationless transitions, we can again ask about the time scales of these "theoretical recurrences [16]". In any real experiment there is some time limit placed upon the length of an experiment, e.g. the length of time we can spend waiting for fluorescence, phosphorescence, etc. This time may be a microsecond, a second, a year, the limits of the patience of a graduate student, etc. Furthermore, the simple model presented in Fig. 1 is only valid for a limited period of time. The molecule has tacitly been assumed to be isolated, but at any nonzero pressure the molecule will collide with another after some time τ_{coll} ($\approx 10^{-4}$ sec at 1μ pressure). Even at "zero pressure," the molecule would collide with the walls of the vessel after some time, τ_{wall}. Furthermore, the states of $\{\phi_l\}$ which are isoenergetic with ϕ_s are vibrationally "hot" and would spontaneously emit infrared radiation [16], (as has in fact been observed by Drent and Kommandeur [28] in biacetyl) thereby leading to a molecule with energy too low to "cross-back" to ϕ_s. ("Isolated molecules" in interstellar space are persistently undergoing these processes.) Estimates of the times for infrared emission give $\tau_{IR} \approx 10^{-5} - 10^{-3}$ sec.

Because of the time limitations on the duration of an experiment and on the validity of the model employed, it is only relevant to consider wether irreversible behavior ensues for times less than or on the order of the time limit τ_{max} [16]. If we find relaxation for times $\leqslant \tau_{max}$, *then we call the phenomenon irreversible.* In this case the discrete manifold $\{\phi_l\}$ is an "effective continuum" on the time scale relevant to the experiment; the $\{\phi_l\}$ then represent a dissipative heat sink. Because the system is considered for only a finite time τ_{max}, all energy levels, except the ground state, must have some "width" in energy ϵ_m as required by the "uncertainty principle"

$$\epsilon_m \gtrsim \hbar/\tau_{max}. \tag{5}$$

This energy uncertainty is introduced by associating with each excited state an additional imaginary energy $-i\epsilon_m$, thereby allowing us to consider whether there is practical irreversibility for times on the order of τ_{max} [16]. This procedure avoids the introduction of unattainable recurrences.[a]

5. Excitation and Decay [16]

In order to obtain criteria for the occurrence of irreversible electronic relaxation, it is necessary to consider the excitation and decay processes whereby the non-radiative decay is detected experimentally. Since the general phenomena associated with radiationless processes are observed regardless of the nature of the exciting light source, e.g. a flash lamp, a laser, or a "pocket ultraviolet flashlight," the basic molecular processes can be adequately exhibited by considering conventional light sources [29]. (Coherent light sources can and should also be considered!) Similarly, monochromatic excitation can be considered, since the superposition principle can be used to account for polychromatic excitation. We therefore assume that a "square pulse" of light is incident upon the system for some time τ. Specifically, we let n_A photons of type A (i.e. frequency ω_A, polarization vector e_A and wave vector k_A) be present at time $t = 0$ when the molecule is initially in the ground state ϕ_0. In the interval $0 < t < \tau$, the system is time independent if the radiation field is described completely quantum mechanically, i.e., we merely count photons. Thus, if the total Hamiltonian for the molecule, the radiation, and their mutual interaction is H, the state of the system is

$$\psi(t) = \exp(-iHt/\hbar) |\phi_0, n_A\rangle \quad 0 < t < \tau. \tag{6}$$

At time τ, we select those molecules which are in some excited molecular state θ_j because they have absorbed one photon of type A, i.e.,

$$\psi_{excited}(\tau) = \sum_j |\theta_j\rangle\langle\theta_j(n-1)_A | \exp(-iH\tau/\hbar) |\phi_0, n_A\rangle. \tag{7}$$

The remaining $(n-1)_A$ photons are removed at time τ and the decay of the excited states is monitored. After τ the system again is described by the time independent quantum mechanical Hamiltonian H, so the state is

$$\psi_{excited}(t+\tau) = \sum_j \exp(-iHt/\hbar) |\theta_j \text{vac}\rangle\langle\theta_j, (n-1)_A | \exp(-iH\tau/\hbar) | \phi_0, n_A\rangle \tag{8}$$

---

[a] The introduction of this imaginary energy contribution can be rigorously included by introducing the interactions leading to the dissipative behavior which introduces τ_{max} and then by considering the effective Hamiltonian for molecule plus radiation which incorporates the net effects of these new decay channels. This imaginary energy implies that all states decay at least with rate $2\epsilon_m/\hbar$, so all states are "annihilated" after times on the order of τ_{max}. They then cannot recur.

where $|vac\rangle$ denotes the state with no photons present. We now ask for the probability $P_{F,A}(t,\tau)$ that the excited molecule emit a photon of type F at a time t after the pulse and fluoresce to a vibrational level, $|\phi_0^v\rangle$ of the ground state. By elementary quantum mechanics, this is

$$P_{F,A}(t,\tau) = |\langle \phi_0^v, 1_F | \psi_{\text{excited}}(t+\tau)\rangle|^2 =$$

$$= |\sum_j \langle \phi_0^v, 1_F | \exp(-iHt/\hbar) | \theta_j, vac\rangle \times \qquad (9)$$

$$\times \langle \theta_j, (n-1)_A | \exp(-iH\tau/\hbar) | \phi_0, n_A\rangle|^2.$$

There are two fundamental reasons for having considered the elementary quantum mechanics (6 – 9). Firstly, no specification has been made as to which choice of molecular basis set (θ_j) is to be taken. *Provided the basis set is complete,* i.e.,

$$\sum_j |\theta_j\rangle\langle\theta_j| \equiv 1, \qquad (10)$$

any choice of basis set could be taken [16]. This is the complimentarity principle of quantum mechanics: all physical observables, e.g., $P_{F,A}(t,\tau)$, must be independent of the choice of a complete basis set [30]. The basis set $\{\theta_j\}$ could be the molecular eigenstates $\{\psi_n\}$, the BO functions $\{\phi_s, \phi_l\}$, with *any choice of the BO approximation,* or any other basis set, provided it is complete for the accessible molecular excited states.

Since nonradiative decay rates $k_{n.r.}$ are physical observables, this general complimentarity principle of quantum mechanics implies that this rate also must be independent of the choice of basis set, e.g., of which form of the BO approximation is employed [31]. However, in practice $k_{n.r.}$ is often taken as the nonradiative decay rate of the "initially prepared" state ϕ_s. The wave function ϕ_s differs slightly for various choices of the BO approximation, so it might be expected that the rates $k_{n.r.}$ obtained from the different BO approximations should not be the same. Since both popular popular versions of the BO approximation provide the same kind of energy level scheme (Fig. 1), if ϕ_s in both of these BO versions are reasonably faithful descriptions of the "initially prepared state ϕ_s," the rates $k_{n.r.}$ calculated from either the crude BO or the adiabatic BO approximations must be essentially identical, provided the full quantum mechanical expression for $k_{n.r.}$ is evaluated! The fact that the calculated decay rate (or width of a resonance) is independent of the choice of the "initially prepared state" (so long as it is close to the true "physical" state) is well known in the case of resonances in electron-atom scattering [32]. This result is obtained, however, only when the full quantum mechanical expression is employed for the decay rate. In the case of radiationless processes in large polyatomic molecules, $k_{n.r.}$ is often expressed in terms of the lowest order nonvanishing "golden rule" rate expression [6-9]. As the "golden rule" rate k_{gold} neglects all higher order corrections, k_{gold} *is not the exact decay rate and need not be the same for different versions of the BO approximation.* Thus, the decay rate $k_{n.r.}$ may be evaluated for any choice of the BO approximation [31]; but when approximations are introduced, it then becomes meaningful to ask which version of the BO approximation will be convenient. This question is considered further in Sect. 9 when the calculation of $k_{\text{n.r.}}$ is discussed.

The second reason to introduce the derivation (6 – 9) is to note that all that is required to evaluate the absorption and emission probability $P_{F,A}(t, \tau)$ of (9) are matrix elements of the evolution operator $\exp(-iHt/\hbar)$. (These matrix elements are the conventional probability amplitudes $a_j(t)$.) When considering a situation in which many different kinds of decay processes are involved, e.g. radiative and nonradiative decay, it is not always convenient to deal directly with the matrix elements of $\exp(-iHt/\hbar)$, the $a_j(t)$. Rather, it is simpler to introduce (imaginary) Laplace transforms [16] in the same manner that electrical engineers use them to solve ac circuit equations [33]. Thus, if E is the transform variable conjugate to t, the transforms of $a_j(t)$ are $g_j(E)$. The quantities $g_j(E)$ can also be labeled by the "initial" state k and are denoded by $G_{jk}(E)$. It is customary in quantum mechanics to collect all these $G_{jk}(E)$ into a matrix $G(E)$. Since matrix methods in quantum mechanics imply some choice of basis set and all physical observables are independent of the chosen basis set, it is convenient to employ operator formulations. If $G(E)$ is the operator whose matrix elements are $G_{jk}(E)$, then it is well known that $G(E)$ is the Green's function [16,30,34] or resolvent operator

$$G(E) = (E - H)^{-1} \tag{11}$$

which satisfies inhomogeneous Schrödinger equation. (1 is the unit operator.)

$$(E - H)\, G(E) = 1. \tag{12}$$

Green's function techniques have been developed to a high degree of sophistication and power in connection with applications to scattering theory [34], etc. [30], and they are beyond the scope of the present review. However, we merely note that by taking the imaginary part of E to be $-\epsilon_m$, we insure that all "unphysical" recurrences for times $\gtrsim \tau_{\text{max}}$ are removed [16]. Thus, these Green's function techniques enable us to treat the simultaneous radiative and nonradiative decay of a large number of closely coupled states [14] in a very compact basis-set-independent manner [16,17].

We briefly note in passing that because of the sum over molecular states $\{\theta_j\}$ in (9), there is *the possibility* that cross-terms may contribute, thereby leading to quantum mechanical interference effects [13-18].

6. Irreversibility Criteria

Before discussing the completely general cases, it is instructive to consider the simple model of Bixon and Jortner [10,13-15]. In this model, the states $\{\phi_l\}$ are taken to be equally spaced in energy, (spacing is ϵ) and the $\nu_{sl} = \nu$ are all the same. The virtue of this model is that it enables exact solutions for the radiative and nonradiative decays. When the radiative lifetime τ_R (l) of the $\{\phi_l\}$ are all $\ll \tau_{max}$, the situation often corresponds to e.g., $S_1 \rightarrow T_1$ intersystem crossing. In this case introducing the energy spread ϵ_m implies that for

$$\tau_{max} \ll \tau_{rec} = \hbar \rho_l = \hbar/\epsilon, \qquad (13)$$

there is irreversible electronic relaxation [15]. In (13) τ_{rec} is the recurrence time which is related to the density of states ρ_l of $\{\phi_l\}$ [10]. All that (13) implies is that if the molecule in ϕ_s "cross-over" to $\{\phi_l\}$ and the experiment is over ($t \gtrsim \tau_{max}$) before the molecule can "cross-back," (τ_{rec} is the time required for "crossing-back") then the radiationless transition is irreversible. On the other hand, if the $\{\phi_l\}$ have appreciable radiative decay rates and

$$\tau_R (l) \ll \tau_{max}, \qquad (14)$$

this situation corresponds to e.g., $S_2 \rightarrow S_1$ internal conversion. Then the irreversibility criterion is (approximately)

$$\tau_R (l) \ll \tau_{rec}, \qquad (15)$$

which states that if the molecule in ϕ_s "crosses-over" to $\{\phi_l\}$, and then it radiates from $\{\phi_l\}$ before it can "cross-back" to ϕ_s, the irreversible electronic relaxation is complete [16]. (15) is even obtained when $\tau_{max} \rightarrow \infty$, or $\epsilon_m \rightarrow 0^+$ in the usual way [34].

For the case of (13), the nonradiative decay rate is given by

$$k_{n.r.} = \frac{2\pi}{\hbar} v^2 \rho_l, \qquad (16)$$

so applying the condition that $k_{n.r.}$ be greater than the recurrence rate leads to the familiar condition

$$v \rho_l \gg 1, \qquad (17)$$

as an irreversibility criterion [1,10,13-15]. The criteria (13) and (15) are, however, more general than (17) for their specific cases of applicability. In the case that $k_{n.r.} \gg \tau_{rec}^{-1} \gg \tau_{max}^{-1} \gg \tau_R^{-1} (l)$ the condition (17) would imply that there is irreversible electronic relaxation, while the correct criterion (13) implies that considerable recurrence behavior is observed instead of electronic relaxation. However, (13) and (15) are special cases of the general irreversibility condition for the above model [16],

$$\tau_{max}^{-1} + \tau_R^{-1} (l) \gg \tau_{rec}^{-1}. \qquad (18)$$

In equation (18) the left hand side is the total rate of the prevention of recurrence via either ending the experiment by collision, say, or radiating to a lower level. We note that it is a trivial matter to include the effects of IR emission in $\{\phi_l\}$ by including $\tau_{IR}(l)$ in $\tau_R(l)$.

In real molecules the $|v_{ls}|^2$ are rapidly varying functions of the particular state ϕ_l. This arises because for a particular ϕ_s, only some $\{\phi_l\}$, call them $\{\phi_b\}$, will have the favorable symmetries such that the Franck-Condon factors between ϕ_s and $\{\phi_b\}$ are appreciable. The remaining and more numerous states of $\{\phi_l\}$, the $\{\phi_w\}$ have unfavorable symmetries and, hence, small Franck-Condon factors. Thus, in general

$$|v_{sb}| \gg |v_{sw}| \tag{19}$$

characterizes real molecules [16,17]. The states $\{\phi_w\}$ and more importantly $\{\phi_b\}$ are not equally spaced, however, (19) is the more severe criticism of the simple equally-spaced, equal-coupling model. In discussing the simplified model, we used the conventional language of "crossing-over" because of its greater familiarity. In the general case, the notion of "crossing-over" loses its utility. Rather than imposing this terminology upon all discussions of radiationless transitions, *it is imperative to let the Schrödinger equation tell us the natural language to be employed* [29,35,36].

7. The Small, Large and Intermediate Molecule Limits

Let the basis set still be the BO states ϕ_s and $\{\phi_l\}$ because, in practice, these are the most readily accessible; they are, in fact, the usual starting points. Since we wish to focus upon all the diverse molecular phenomena which are classified as involving radiationless processes, it is necessary to center attention upon the molecule. This focus is best obtained by considering the effective Hamiltonian H_{eff}, for the molecule which accounts for all relaxation mechanisms other than the intramolecular nonradiative decay. (The use of effective Hamiltonians is popular in considering the relaxation processes associated with studies of magnetic resonance [37].) For the present case, the effective Hamiltonian is [16,17]

$$H_{eff} - i\epsilon_m = H_M - \frac{1}{2}i\hbar\Gamma - i\epsilon_m, \tag{20}$$

where H_M is the molecular Hamiltonian in the absence of radiation (i.e., in the absence of spontaneous emission), ϵ_m gives each level an uncertainly width so that practical irreversibility on the time scale τ_{max} can be investigated, and Γ is the radiative damping matrix. The latter is a generalization of the "golden rule" radiative decay rates, and is defined by its matrix elements in an arbitrary basis [16-18],

$$\Gamma_{nm} = \frac{2\pi}{\hbar} \sum_\alpha \sum_e \int d\Omega_k \langle n|H_{int}|\alpha, ke\rangle$$
$$\times \langle \alpha, ke|H_{int}|m\rangle \rho_{photon}(\omega), \tag{21}$$

where H_{int} represents the interaction of matter and radiation, and the summation is over all lower states α to which the states m and n may spontaneously radiate and all polarization directions e. The integration, $\int d\Omega_k$, is over all directions of photon propagation, and $\rho_{photon}(\omega)$ is the density of photon states at the frequency of the emitted photon $\omega = c|k|$, where c is the velocity of light. In the BO basis we usually consider that Γ is diagonal. Consider the case in which ϕ_s has a much larger radiative decay rate $\Gamma_s \equiv \Gamma_{ss}$ than the $\{\phi_l\}$. This case would correspond to ϕ_s, e.g., S_1 and $\{\phi_l\}$ as T_1, where the IR fluorescence and phosphorescence rates of T_1 are often quite negligible in comparison to Γ_s. [The general case in which the $\{\Gamma_{ll}\}$ are not negligible with respect to Γ_s can easily be considered.] Thus, in the BO basis, the matrices H_M and Γ have the forms

$$H_M = \begin{pmatrix} E_s \overset{\leftarrow}{} v_{sl} \overset{\rightarrow}{} \\ \uparrow \quad \nwarrow \quad O \\ v_{ls} \quad E_l \approx E_s \\ \downarrow \quad O \quad \searrow \end{pmatrix} \qquad \Gamma = \begin{pmatrix} \Gamma_s & O \\ O & \bigcirc \end{pmatrix} \qquad (22)$$

i.e. in H_M there is coupling v_{ls} between ϕ_s and $\{\phi_l\}$, while Γ has only one non-zero matrix element. The molecular eigenstates $\{\psi_n\}$ are obtained by diagonalizing H_m; but that transformation scrambles up Γ, leaving it with many non-zero off-diagonal matrix elements. Nonzero Γ_{nm}, $n \neq m$, imply that there is the physical possibiliy of having a *virtual two photon process* in which a molecule initially in state n emits a photon of energy $\hbar\omega$ in making a transition to some lower molecular state α. *Then the molecule absorbs* the same photon and is excited from α to state m. Thus, the net effect is a transition from molecular state n to molecular state m, with no change in the number, frequency, etc., of photons present. These virtual two photon processes which are represented schematically in Fig. 3, are usually highly unlikely, except when the two states n and m are nearly degenerate. As stressed earlier, when states are degenerate, interactions between them have important qualitative implications, no matter how small the interactions. Thus, these virtual processes are important when

$$\hbar\Gamma_{nm} \gtrsim |E_n - E_m|, \qquad (23)$$

where E_n and E_m are the energies of the states n and m.

In the small molecule limit, except in the case of accidental degeneracies, the energy level spacings are much greater than any radiative widths $\hbar\Gamma_{nm}$. Hence the *molecular eigenstates* $\{\psi_n\}$ *diagonalize the effective Hamiltonian (20) for the case of (22) to a very good approximation in the small molecule limit* [16] Each molecular eigenstate ψ_n contains a fraction of the total oscillator strength f of ϕ_s to the extent that it contains the state ϕ_s, i.e., $|\langle \psi_n | \phi_s \rangle|^2$. Thus, each molecular eigenstate has a longer radiative lifetime $\tau_R(n)$ than the radiative lifetime Γ_s^{-1} that would be deduced from the integrated absorption intensity,

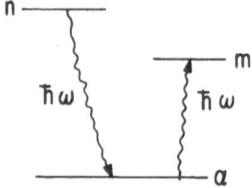

Fig. 3. The schematic representation of a virtual two photon process which is allowed when there is a nonzero matrix element of the radiative damping matrix Γ_{nm} connecting two states n and m. This process is negligible unless n and m are nearly degenerate

$$\tau_R(n) > \Gamma_s^{-1} \equiv \tau_R(s). \qquad (24)$$

The state ϕ_s is not really a physically accessible state, since in order to prepare the molecule initially in ϕ_s, it would be necessary to have exciting radiation with particular coherence between different frequencies which correspond to all the observed spectral lines from the molecular states which contain this zeroth order state ϕ_s to the initial state ϕ_0. The molecular eigenstates should then have unit quantum yields at very low pressure, a result which is not inconsistent with experimental extrapolations to zero pressure. This small molecule limit is depicted schematically in Fig. 4a.

In the large molecule, statistical, limit, there would be a large number of molecular eigenstates which satisfy the condition (23), making the virtual two photon processes of Fig. 3 important for a very large number of states [14, 16, 18]. An interesting basis set to employ might be the states $\{\chi_r\}$ which diagonalize H_{eff} of (20). For the case that the $\{\phi_l\}$ form a dense enough set that the distribution of recurrence times τ_{rec} are all $\gg \tau_{max}$, it is simpler to consider the resonant states which closely approximate the simple physical behavior of the "initial" state ϕ_s which can decay radiatively and nonradiatively. The $\{\phi_l\}$ form an effective continuum into which ϕ_s may nonradiatively decay on the time scales of relevance to real experiments. Thus, ϕ_s may decay into $\{\phi_l\}$ and also decay radiatively, so its observed radiative lifetime and quantum yield should correspond to (1) and (2). The "state" ϕ_s is also very close to the "physical state" which is accessible upon excitation. The statistical limit is presented pictorially in Fig. 4b. It should also be noted that the conclusion of the Bixon-Jortner model that the absorption lineshape to ϕ_s be Lorentzian in the statistical limit is no longer valid in the real and general case when the v_{ls} are rapidly varying.

With a large and small molecule limit, there must also be an intermediate case. There are a number of possible manifestations of the intermediate case, so one interesting choice is discussed [16]: There may be situations in which the weakly coupled states $\{\phi_w\}$ are sufficiently dense that on the time scales of real experiments, the $\{\phi_w\}$ behave as an effective continuum for decay of the zeroth

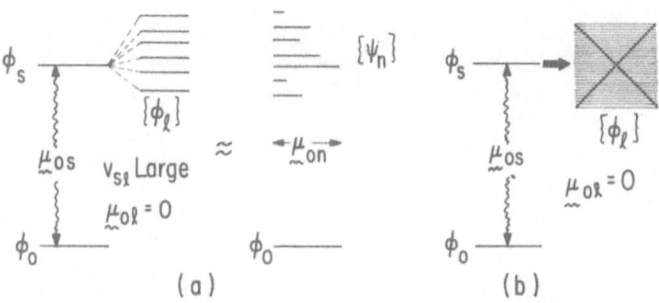

Fig. 4. (a) A schematic representation of the small molecule limit. The states are the same as those represented in Fig. 1. The molecular eigenstates approximately diagonalize the effective molecular Hamiltonian (20), and each carries only a portion of the original oscillator strength to ϕ_0. (b) A representation of the statistical limit. The $\{\phi_l\}$ form a dense manifold of states which acts as a dissipative quasicontinuum on the time scales of real experiments. ϕ_s can therefore decay radiatively to ϕ_0 and nonradiatively to $\{\phi_l\}$

order state ϕ_s. (This does not imply that ϕ_s is a physically accessible state!) ϕ_s is also strongly coupled to the sparser distribution of states $\{\phi_b\}$ which do not act as a dissipative sink. The set of zeroth order states $\{\phi_s, \phi_b\}$ represent a set of closely coupled states *which can decay into two continua:* ϕ_s is coupled to the vibronic continuum of $\{\phi_w\}$ and to the radiative decay channel. (The zeroth order $\{\phi_b\}$ could also have non-negligible radiative decay rates.) Thus, the states $\{\phi_s, \phi_b\}$ appear to be just like the small molecule limit, except that there is the additional continuum formed by the vibronic manifold $\{\phi_w\}$. In this case, we could consider the effective Hamiltonian H_{sb} for the set of levels $\{\phi_s, \phi_b\}$ which can be shown to be [16)]

$$H_{sb} = H_M(sb) - \frac{1}{2} i\hbar \Gamma(sb) - \frac{i\hbar}{2}\Delta. \qquad (25)$$

$H_M(sb)$ and $\Gamma(sb)$ are the molecular Hamiltonian and the damping matrix, respectively, within the states $\{\phi_s, \phi_b\}$, while Δ is the matrix of nonradiative decay rates which has only one nonzero matrix element $\Delta_s = \Delta_{ss}$ within this BO basis set. The states $\{\chi_r\}$ which diagonalize (25) are then the resonant states, each with some energy E_r and some radiative and nonradiative decay rates Γ_r and Δ_r, respectively. When these resonances are overlapping, i.e.,

$$\frac{\hbar}{2}(\Gamma_r + \Delta_r + \Gamma_{r'} + \Delta_{r'}) \gtrsim |E_r - E_{r'}|, \qquad (26)$$

purely quantum mechanical interference effects may be observed. The intermediate case discussed above is represented schematically in Fig. 5 along with a qualitative spectrum. In the spectrum there is a broad background arising from the $\{\phi_w\}$ and sharper resonances superimposed upon this background. (The re-

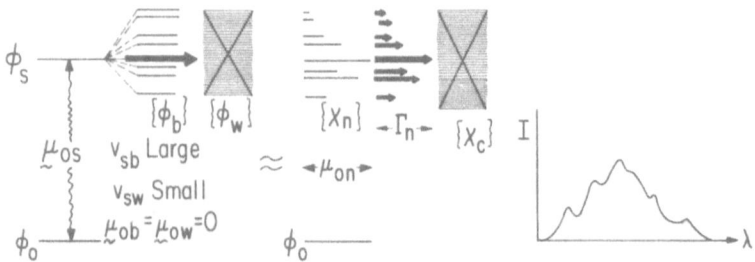

Fig. 5. One particular realization of the intermediate case is depicted showing the strongly and weakly coupled states $\{\phi_b\}$ and $\{\phi_w\}$, respectively. The latter are assumed to form an effective quasi-continuum. The states $\{\chi_r\}$ which diagonalize the effective Hamiltonian (25) are the resonant states, each of which has both a radiative and nonradiative decay rate. A schematic representation of the absorption spectrum is presented for this particular type of intermediate case molecule where it has been assumed that there is no overlapping of resonances

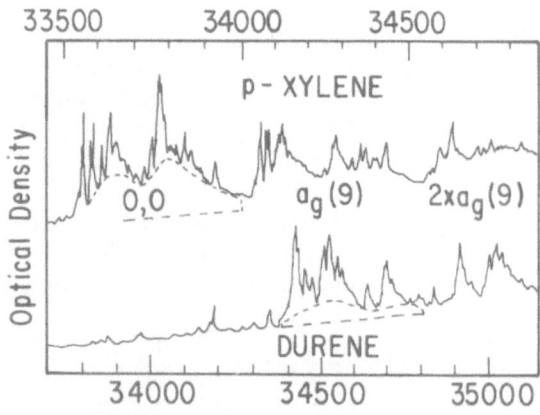

Fig. 6. The $^1B_{2u}{}^+(S_2)$ absorption spectra of naphthalene in p-xylene and durene at 2.2 °K obtained by Wessel and McClure is reproduced from their abstract [38]. The energy scales have been shifted so the $S_1(^1B_{3u}^-)$ origins coincide

sonances are depicted here as being nonoverlapping.) Thus, it is expected that the quantum yields (which are less than unity) should vary as we pass through the broad background and all the resonances, and the observed radiative decay lifetimes will also vary.

The second singlet ($^1B_{2u}^+$) absorption spectra of naphthalene in p-xylene and durene at 2.2 °K obtained by Wessel and McClure represent a good example of the intermediate case [38]. The spectra are reproduced in Fig. 6, where the energy scales have been shifted so the $S_1(^1B_{3u}^-)$ origins coincide in both matrices.

The $S_2 - S_1$ energy gap ΔE is only ≈ 3000 cm^{-1}, so the S_1 vibrations in the energy region corresponding to the $0 - 0$, the $a_g(9)$, etc., bands in S_2 can readily be enumerated. The symmetry of $B_{3u} \times B_{2u}$ is b_{1g}, and the structure in the $0 - 0$ and the $a_g(9)$ region can be assigned to interactions between the S_2 vibronic state with vibronic states of S_1 which contain one quantum of b_{1g} vibrations and some number of a_g vibrations [38].

8. Vibrational Relaxation and Nonradiative Decay

The results of the previous section can immediately be used to obtain an understanding of the effects of vibrational relaxation upon the radiationless processes in dilute gases. When a statistical limit molecule in the vibrationless state of ϕ_s collides with the buffer gas, it is expected that its quenching rate would be much less than the rate of vibrational relaxation of the vibrationally hot states $\{\phi_l\}$. Thus, to zeroth order in low pressures, quenching of ϕ_s is neglected relative to the rapid vibrational relaxation of $\{\phi_l\}$. Since the $\{\phi_l\}$ act as a dissipative sink to ϕ_s, once a molecule "crosses-over" from ϕ_s to $\{\phi_l\}$, it cannot "cross-back" even when the molecule is at essentially zero pressure. Thus, the vibrational relaxation of $\{\phi_l\}$ should not affect the nonradiative decay of statistical limit molecules (in their vibrationless state). This conclusion is consistent with the observed constancy of fluorescence lifetimes and quantum yields of statistical limit molecules. For anthracene $k_{n.r.}$ is unchanged even when the molecule is in an inert hydrocarbon solvent. Thus, the quenching rate of the vibrationless component of S_1 of anthracene must be small compared to τ_f^{-1}, $k_{n.r.}$, etc.

In the small molecule limit, the physical situation corresponds more closely to that described by the molecular eigenstates $\{\psi_n\}$. Each molecular eigenstate contains some admixture of the zeroth order states, the vibrationally "cold" ϕ_s (when we consider excitation to the lowest states) and the vibrationally "hot" $\{\phi_l\}$. Thus, each molecular eigenstate should undergo rapid vibrational relaxation to the extent that it contains $\{\phi_l\}$ in agreement with experiment. Since an eigenstate which contains a large part of $\{\phi_l\}$ only has a small component of ϕ_s and the original oscillator strength, qualitatively it is expected that the molecular eigenstates with the longer τ_f should have larger vibrational relaxation rates. (The real situation is, of course, complicated by the fact that neighboring zero order states $\{\phi_l\}$ may have very different excited vibrations, and consequently differing vibrational relaxation rates.)

In the intermediate case the $\{\phi_w\}$ are already dissipative, so their rapid vibrational relaxation should not affect the nonradiative decay rates. However, each of the resonant states χ_r contain some of the zeroth order states ϕ_s and ϕ_b, and consequently they should exhibit vibrational relaxation to the extent that they contain $\{\phi_b\}$ (for ϕ_s the vibrationless state). At higher pressures at which the vibrational relaxation rates of $\{\phi_b\}$ become greater than the averages of τ_f^{-1} and of $k_{n.r.}$ for the resonant states, this pressure broadening will effectively lead to $\{\phi_b\}$ being an effective quasicontinuum, i.e., being a dissipative sink. Thus, a molecule which corresponds to the intermediate case at zero pressure becomes transformed to the statistical limit at higher pressures. This implies that the quantum yield should decrease with increasing pressure until it reaches a limiting value which is independent of pressure in "inert" media (those which only lead to vibrational relaxation). This situation corresponds to the observed benzene S_1 fluorescence quantum yields as noted in Sect. 2. Thus, on the basis of this pressure dependence, benzene can be characterized as an intermediate case molecule.

The above discussion represents a satisfactory qualitative description of the effects of vibrational relaxation upon radiative decay rates. When ϕ_s is a excited vibronic state, it can contribute directly to the vibrational relaxation. It may also be necessary in some cases to consider direct collision induced "crossings." The collision induced transitions between the various components of ϕ_s itself are, of course, of great interst.

9. Nonradiative Decay Rates and the Choice of the Born-Oppenheimer Approximation

There has been considerable controversy in the literature as to whether the nonradiative decay rates $k_{n.r.}$ are to be evaluated using the adiabatic Born-Oppenheimer (ABO) or the crude Born-Oppenheimer (CBO) approximation [8, 10, 31, 39-43]. In Sect. 5 it was noted that the complimentary principle of quantum mechanics requires that the rates exactly calculated within these two schemes be the same provided that ϕ_s in both schemes is, as expected, a reasonable approximation to the true physical state. As noted also, in both the ABO and the CBO approximations. we have the same mechanistic schemes of ϕ_s coupled to the effective quasicontinuum $\{\phi_l\}$. Thus, both cases represent differing, yet reasonable representations of ϕ_s and $\{\phi_l\}$. In the present discussion it is necessary to consider the remainder of the vibronic states of the moleculae $\{\phi_c\}$ in addition to $\{\phi_s, \phi_l\}$. Thus, the coupling matrix elements are no longer the "effective" values, but are actual matrix elements of the perturbing Hamiltonian.

The Green's function formalism alluded to in Sect. 5 can be employed to provide general expressions for $k_{n.r.}(s)$ in the statistical limit as [16,17,44]

$$k_{n.r.}(s) = \frac{2\pi}{\hbar} \sum_l |\langle s | R_s(E_s) | l \rangle|^2 \, \delta(E_s - E_l), \qquad (27)$$

where E_s is the exact (shifted) energy of ϕ_s. Since $k_{n.r.}(s)$ is a number, the Dirac delta function requiring conservation of energy is to be taken as a smooth representation of a delta function, reflecting the fact that electronic relaxation occurs on the time scales which correspond to real experiments [39]. The summation in (27) is over all vibronic components in $\{\phi_l\}$, and *the matrix elements of* R_s *contain terms to all orders of perturbation theory* [44],

$$\langle s \,|R_s(E_s)\,|l\rangle = \langle \phi_s \,|\,V\,|\phi_l\rangle + \sum_{c \neq s} \frac{\langle \phi_s \,|V\,|\phi_c\rangle\langle \phi_c \,|V\,|\phi_l\rangle}{E_s - E_c^0} +$$

$$+ \sum_{c,c' \neq s} \frac{\langle \phi_s \,|V\,|\phi_c\rangle\langle \phi_c \,|V\,|\phi_{c'}\rangle\langle \phi_{c'} \,|V\,|\phi_l\rangle}{(E_s - E_c^0)\,(E_s - E_{c'}^0)} + \ldots , \qquad (28)$$

where the summation over $c, c', \ldots$ may include the $\{\phi_l\}$. In (28) the energies E_c^0 are those corresponding to the eigenstates ϕ_c of the zero order molecular Hamiltonian H_0. The perturbation V is the difference between the true molecular Hamiltonian H_M and the approximation H_0,

$$H_M = H_0 + V. \qquad (29)$$

Thus, (27–29) can be written for both the CBO and the ABO approximations, and the calculated rates, including terms to all orders, should be identical in both cases.

In practive, however, only the lowest order nonvanshing contributions in (28) are considered. If these terms are then taken as the effective interactions v_{sl}, the approximate "golden-rule" rate is written as

$$k_{gold} = \frac{2\pi}{\hbar} \sum_l |v_{sl}|^2 \,\delta(E_s - E_l), \qquad (30)$$

which is the conventional form. *Because (30) is an approximation, there is no longer the requirement that* k_{gold} *calculated by the ABO or the CBO agree, or in fact coincide with the exact* $k_{n.r.}$ *of (27).* In fact, a situation may arise in which terms in (27) may be required to all orders of perturbation theory for some intermediate states (possibly from the $\{\phi_l\}$). The rates calculated from (27) would in principle still be correct, but the physical situation would correspond to a Jahn-Teller effect (usually dynamical) in conformity with Teller's suggestion [43]. However, often the true series (27) will converge in finite order. There are then a few questions to be posed:

(a) Which parts of the perturbation V provide the dominant contributions to (27)?

(b) What are the magnitudes of the corrections to (30) within the CBO and ABO approximations? Also, which approximation scheme gives the smallest corrections, so it would lead to faster convergence?

(c) Finally, when are the "golden-rule" rates (30) virtually identical to the exact rates (27)? Does this equality occur for one or both of the CBO and/or the ABO?

Clearly, there is a lot of room for further investigation of these questions, even with the difficulties in obtaining the ABO or CBO molecular wave functions. A few points can, however, be illuminated at present.

Burland and Robinson [41] have presented semi-quantitative arguments (in the case of internal conversion) to determine the ratios of the electronic matrix elements β in what they call the Herzberg-Teller scheme and the ABO approximation. Their arguments may be qualtitativeley correct only if the CBO approximation is employed instead of the Herzberg-Teller scheme[b].

Accepting their result in the CBO case, they claim that the ratio of the electronic coupling matrix element β_{ABO} in the ABO to β_{CBO} in the CBO approxiamtions is

$$\frac{\beta_{ABO}}{\beta_{CBO}} \approx 0.1. \tag{31}$$

Even if this result (31) is qualitatively correct, however, their conclusion that (31) implies that the CBO coupling is more efficient in inducing nonradiative decay is wrong. The result (31) implies that the higher order corrections k_{cor} in the CBO are in fact larger than those in the ABO, since

$$\frac{(k_{cor}/k_{gold})_{CBO}}{(k_{cor}/k_{gold})_{ABO}} \approx \frac{(\beta/\Delta E)_{CBO}}{(\beta/\Delta E)_{ABO}} \approx \frac{\beta_{CBO}}{\beta_{ABO}} \approx 10. \tag{32}$$

Equation (32) leads to three possibilities: (i) $(k_{gold})_{CBO}$ and $(k_{gold})_{ABO}$ are essentially equivalent and exact, (ii) both are incorrect since the higher order corrections are non-negligible, or finally, (iii) $(k_{gold})_{ABO}$ has negligible higher order corrections and is exact but $(k_{gold})_{CBO}$ is not the true rate. Since the ABO involves a diagonalization of a larger part of the molecular Hamiltonian, the ABO rate expressions should converge more rapidly than the CBO ones. Lefebvre has argued that the CBO states are not physically accessible [43] in the same sense that the $\{\phi_s, \phi_l\}$ were considered in Sect. 7 for the small molecule limit.

However, in the absence of accurate ABO and CBO electronic wave functions, the use of the CBO "golden-rule" rate expression is extremely useful in mechanistic discussions of question (a) above and in providing semi-quantitative results. This use of the CBO employs the basic symmetry properties of the molecular states, i.e., the essential feature of Herzberg-Teller theory. It should be noted in passing that when considering the true rates (27), the matrix elements can be considered to be those of an operator $R_s(E_s, q, Q)$ which depends upon both electronic (q) and nuclear (Q) coordinates in the CBO and the ABO. We can generalize the Franck-Condon principle to write matrix elements as a product of an electronic factor and a Franck-Condon factor (which includes the effects of the promoting mode as discussed in the next section), where the elec-

---

b) In the latter case the energy denominators are Q-dependent (see Eqs. (33 - 35) below), while in the former they are not. The complete neglect of the Q-dependence is incorrect in the latter case.

tronic factor is to be evaluated at some "Q-centroid" for the nuclear coordinates [31]) by analogy with the r-centroid method used for diatomic molecules [45]).

Although it was not necessary for the above development, the CBO and ABO approximations should be briefly reviewed [46]). In both approximations the vibronic wave function $\phi^{BO}(q, Q)$ is written as a product of an electronic wave function $\psi(q, Q)$ and a nuclear vibrational wave function $\chi(Q)$. In the CBO), the electronic functions are taken to be independent of Q (they are all evaluated at some Q_0), while in the ABO the ψ are evaluated parametrically for each value of Q. Thus, the vibronic functions are

$$\phi^{CBO}(q, Q) = \psi^{CBO}(q, Q_0)\, \chi^{CBO}(Q) \tag{33a}$$

$$\phi^{ABO}(q, Q) = \psi^{ABO}(q, Q)\, \chi^{ABO}(Q). \tag{33b}$$

The exact wave function can always be represented as a superposition of CBO or ABO functions. $\psi^{CBO}(q, Q_0)$ is an eigenfunction of the electronic Hamiltonian, $H_{el}(q, Q_0)$, at a particular Q_0, while $\psi^{ABO}(q, Q)$ is an eigenfunction of $H_{el}(q, Q)$. Here $H_{el}(q, Q) = T(q) + V(q, Q)$, where $T(q)$ is the kinetic energy of the electrons and $V(q, Q)$ the electron-nuclear Coulomb energy. In general H_{el} omits the spin-orbit interaction $H_{so}(q, Q)$, so $H_{so}(q, Q)$ forms part of the perturbation in both the CBO and ABO schemes. The vibrational wave functions $\chi_{j\alpha}(Q)$ for electronic state j are then eigenfunctions of the Hamiltonians (α denotes the vibrational component of j)

$$H_{Nj}^{CBO} = T_N + \langle \psi_j^{CBO}(q, Q_0)|H_{el}(q, Q)|\psi_j^{CBO}(q, Q_0)\rangle_q, \tag{34a}$$

$$H_{Nj}^{ABO} = \langle \psi_j^{ABO}(q, Q)|H_{el}(q, Q) + T_N|\psi_j^{ABO}(q, Q)\rangle_q, \tag{34b}$$

where T_N is the nuclear kinetic energy operator and the subscript q denotes integration only over the electronic coordinates.

The perturbations V are given by

$$\langle \phi_{j\alpha}^{CBO}|V^{CBO}|\phi_{k\beta}^{CBO}\rangle = \langle \phi_{j\alpha}^{CBO}|H_{so}|\phi_{k\beta}^{CBO}\rangle +$$

$$+ \langle \phi_{j\alpha}^{CBO}|T_N + H_{el}|\phi_{k\beta}^{CBO}\rangle(1 - \delta_{jk}) \tag{35a}$$

$$\langle \phi_{j\alpha}^{ABO}|V^{ABO}|\phi_{k\beta}^{ABO}\rangle = \langle \phi_{j\alpha}^{ABO}|H_{so}|\phi_{k\beta}^{ABO}\rangle +$$

$$+ \langle \phi_{j\alpha}^{ABO}|T_N|\phi_{k\beta}^{ABO}\rangle(1 - \delta_{jk}). \tag{35b}$$

These perturbations are to be used in the exact rate expression (27–28) with the zero order energies $E_{j\alpha}^{CBO}$ and $E_{j\alpha}^{ABO}$, respectively.

Within the CBO it is often customary to consider power series expansions such as

$$H_{so}(q, Q) = H_{so}(q, Q_0) + \left(\frac{\partial H_{so}(q, Q)}{\partial Q}\right)_{Q_0} (Q - Q_0) + \dots, \quad (36)$$

and only the constant and linear term in $(Q - Q_0)$ are retained. The CBO is also often developed by considering similar power series expansions of all ABO expressions. The latter is the approach followed by Henry and Siebrand in their analysis of intersystem crossing rates in aromatic hydrocarbons [8,10,42]. They indicate three mechanisms leading to intersystem crossing [42]. (a) Direct spin-orbit crossing due to a nonvanishing $\langle \phi_s^{CBO} | H_{so}(q, Q_0) | \phi_l^{CBO} \rangle$. (b) Vibronic induced spin-orbit crossing (Herzberg-teller mechanism) when

$$\langle \phi_s^{CBO} | \frac{\partial H_{so}}{\partial Q} \right)_{Q_0} (Q - Q_0) | \phi_l^{CBO} \rangle + \langle \chi_{sj}(Q) \left(\frac{\partial \psi_s^{ABO}}{\partial Q}\right)_{Q_0} (Q - Q_0) | H_{so}(q, Q_0)|$$

$$\times \phi_l^{CBO} \rangle + \langle \phi_s^{CBO} | H_{so}(q, Q_0) | \chi_{lj} \left(\frac{\partial \psi_l^{ABO}}{\partial Q}\right)_{Q_0} (Q - Q_0) \rangle \neq 0. \quad (37)$$

(c) Adiabatic coupling when

$$\sum_{c \neq s} \frac{\langle \phi_s^{CBO} | T_N | \phi_c^{ABO} \rangle|_{CBO} \langle \phi_c^{CBO} | H_{so}(q, Q_0) | \phi_l^{CBO} \rangle}{E_s^{CBO} - E_c^{CBO}}$$

$$+ \sum_{c' \neq s} \frac{\langle \phi_s^{CBO} | H_{so}(q, Q_0) | \phi_{c'}^{CBO} \rangle \langle \phi_{c'}^{CBO} | T_N | \phi_l^{ABO} \rangle|_{CBO}}{E_s^{CBO} - E_{c'}^{CBO}} \neq 0, \quad (38)$$

where $T_N \phi_c^{ABO} |_{CBO}$ implies that all derivatives of ψ_c^{ABO} with respect to nuclear coordinates are to be taken at Q_0. We note that they have omitted the possibility of cross terms between mechanisms (a), (b), and (c), which in general must be considered even in lowest order. The mechanisms (a) and (b) both correspond to direct spin-orbit crossing in the ABO. Terms of the form of (38) occur directly within the CBO as third and higher order contributions of (27).

10. Evaluation of Radiationless Decay Rates

The generalized Franck-Condon principle as described in the previous section is assumed to be applicable so the rate (27) may be expressed in the "golden-rule" like form (assuming a single promoting mode k)

$$k_{n.r.}(sj) = \frac{2\pi}{\hbar} |C_{sl}^k|^2 \sum_l |\langle \chi_{si} | \frac{\hbar}{i} \frac{\partial}{\partial Q_k} | \chi_{lj} \rangle|^2 \delta(E_{si} - E_{lj}), \quad (39)$$

where $|C_{sl}^k|^2$ is the electronic factor which is evaluated at the "Q-centroid" $\bar{Q}$ [31], the vibrational components of the electronic states s and l are i and j, respectively, and the energies of the vibronic components are E_{si} and E_{lj}. In writing (39) the electronic wave functions ψ have been implicitly assumed to depend weakly enough on the nuclear positions so that all second derivatives with respect to Q_k can be neglected (Condon approximation). The electronic factors contain terms to all orders of perturbation theory (28), while the remainder of the expression (39) is just the Franck-Condon factors with the constraint of conservation of energy imposed by the Dirac delta function. An ABO basis set is assumed to be taken and the direct spin orbit mechanism is assumed to be zero. [If this is nonzero, the $\frac{\hbar}{i} \frac{\partial}{\partial Qk}$ in (39) is absent. The development in this case simply follows as in the case of (39). If a CBO formulation were employed the operator Q_k would replace $\frac{\hbar}{i} \frac{\partial}{\partial Q_k}$; however, these two operators have the same selection rules, so the results are qualitatively the same.] The modes k for which the electronic factor $C_{ls}^k \neq 0$ are called *promoting modes*. For the case that the molecules are in solution, because of rapid vibrational relaxation, the states si are taken to be distributed according to a Boltzmann distribution $p(si)$. Thus

$$k_{n.r.}(s \to l) = \frac{2\pi}{\hbar} |C_{sl}^k|^2 \sum_{i,j} p(si) |\langle \chi_{si} | \frac{\hbar}{i} \frac{\partial}{\partial Q_k} | \chi_{lj} \rangle|^2 \delta(E_{si} - E_{lj}) \qquad (40)$$

is the rate in "inert" solutions which do not alter the purely intramolecular character of the nonradiative decay.

First consider the case of low temperatures in which si can be taken as the vibrationless component of the electronic state ϕ_s, i.e. the Boltzmann factors $p(si)$ are negligible for all but the lowest vibrational level. It has been popular to make the crude factorization of (39) into a product of the electronic factor, an "average" Franck-Condon factor $\langle F^2 \rangle$, and an "effective" density of states $\rho_l(\Delta E)$ [6,7,9,21,22,47],

$$k_{n.r.}(s \to l) = \frac{2\pi}{\hbar} |C_{sl}|^2 \langle F^2 \rangle \rho_l(\Delta E). \qquad (41)$$

Unfortunately, neither the "averaging" procedure to obtain $\langle F^2 \rangle$ nor the manner in which $\rho_l(\Delta E)$ is to be evaluated are unambiguously defined. Thus, (41) replaces a single experimental quantity by two undefined parameters and an electronic factor which is difficult to evaluate.

Other workers have explicitly evaluated the Franck-Condon factors which appear in (39), taking the maximum term as a measure of the rate [6,9,21,22,47]. These studies have usually been limited to a study of only two vibrational modes. For the case in which the electronic potentials in states l and s are both assumed to be harmonic with the same frequencies and normal modes in both states, the Franck-Condon factors are simply written, and the rate is [6]

$$k_{n.r.}(s \to l) \propto \sum_{\{n_\alpha\}} \left[\prod_\alpha \frac{(x_\alpha)^{n_\alpha}}{n_\alpha!} \right] \delta(\Delta E_k - \sum_\alpha \hbar \omega_\alpha n_\alpha). \qquad (42)$$

In (42) $x_\alpha = \frac{1}{2} \Delta_\alpha^2$ and Δ_α is the dimensionless shift in equilibrium position of the oscillator α between the two electronic states l and s. The oscillators have quanta of energies $\hbar\omega_\alpha$. The promoting mode is assumed to have $\Delta_k = 0$, so the effective energy gap is

$$\Delta E_k = \Delta E - \hbar\omega_k, \tag{43}$$

since the promoting mode which has zero quanta in the vibrationless component of s must gain one quantum in going to l (because of the selection rules on $\frac{\partial}{\partial Q_k}$). The summation in (42) is over all possible vibrational states in l, and only the contribution to $k_{n.r.}(s \to l)$ arising from a single promoting mode is explicitly exhibited. We could consider the direct numerical calculation of (42), or investigate other theoretical approaches to its evaluation. However, there is no need for all that effort since expressions identical to (42) have been extensively studies in other contexts. Note that the individual Franck-Condon factors $(x_\alpha)^{n_\alpha}/n_\alpha!$ look identical to Boltzmann counting weights which give the number of ways of putting n_α indistinguishable particles into x_α boxes. Thus, the expression (42) has precisely the form of the partition function for a set of Boltzmann particles with "degeneracies" x_α and energies $\hbar\omega_\alpha$. The delta function in (42) implies that the total energy is fixed at ΔE_k, so (42) corresponds to the partition function in a microcanonical ensemble (fixed total energy) [48]. There is no restriction on the total number of "particles"

$$N = \sum_\alpha n_\alpha, \tag{44}$$

which is appropriate to vibrational quanta. This analogy between the individual Franck-Condon factors and Boltzmann factors immediately provides us with some well studied approximate methods for evaluating (42). In the statistical mechanical analogy of (42), when ΔE_k is large — this corresponds to the statistical limit in the radiationless decay process — (42) is approximated by taking the maximum term in the summation with the energy conservation $\Delta E_k = \sum_\alpha n_\alpha \hbar\omega_\alpha$ introduced as a constraint by the method of Lagrange multipliers [48]. Using Stirling's approximation, the result can be shown to give [23,39]

$$n_\alpha^0 = x_\alpha \exp(b \hbar\omega_\alpha), \tag{45}$$

where n_α^0 is the value of n_α for the maximum of (42), (i.e., n_α^0 is the most probable number of quanta in the αth vibration in the final electronic state) and $b > 0$ is the Lagrange multiplier. (b looks like an inverse temperature!) Note that (45) implies that the modes of maximum frequency, for which $x_\alpha \neq 0$, accept the most quanta and hence contribute most to the rate. In aromatic hydrocarbons these are the $C-H$ stretches which are known to dominate the nonradiative rate, i.e., the CH vibrations are "the best accepting modes." The maximum term can be written as [23,39]

$$k_{n.r.} (s \to l) \propto \exp\left[-\gamma \Delta E_k / \hbar \omega_M \right], \qquad (46)$$

where γ is in the range $1 - 3$ for $T_1 \to S_0$ intersystem crossing in aromatic hydrocarbons. ω_M corresponds to the maximum molecular frequency, and, by simply substituting (45) into the energy conservation condition and solving for b, γ can be explicitly expressed as a function of x_M, ω_M, ΔE_k and the other modes of the molecule. For $T_1 \to S_0$ transitions in aromatic hydrocarbons the C$-$C skeletal stretches are expected to have a measurable effect upon γ in (46) [39]. For aromatic hydrocarbons $\omega_M \approx 3100$ cm^{-1}, while for the perdeutero compounds $\omega_M \approx 2300$ cm^{-1}. Thus, (46) explicitly exhibits the energy gap law dependence and deuterium isotope effect that were found empirically by Siebrand (Fig. 2) [6,22]. The energy gap law was first derived by Englman and Jortner for the case in which all modes except the dominant C$-$H stretches are ignored [49]. (They used a different method which is discussed below.) Therefore, by evaluating the maximum term in (42) algebraically instead of numerically, the energy gap law is simply and explicitly shown to follow. In (42) and (46) any number of modes can be considered as opposed to the numerical calculations in which only two modes were treated. The rate (42) can be generalized to cover cases in which a particular excited vibronic component of s is initially excited by narrow band monochromatic radiation [23]. This generalization was employed to evaluate the dependence of $k_{n.r.}$ upon the initial vibronic state for $S_1 \to T_1$ intersystem crossing in benzene (see Sect. 12) [23]. However, the Boltzmann statistics method is only intended to be semi-quantative since it cannot account for the important pre-exponential factors, which would make (46) an equality, and it does not allow for frequency shifts in the modes between s and l.

Before considering the more general methods which provide the full rates, it is interest to reflect upon the results obtained by this simple approximation. Because all the $x_\alpha < 1$ in aromatic hydrocarbons, all the Franck-Condon factors for $n_\alpha \neq 0$ are unfavorable compared to the case of $n_\alpha = 0$. Thus, the molecule would like to keep all the $\{n_\alpha\}$ a minimum and, hence, N at a minimum. It can accomplish this by putting the available energy ΔE_k into the modes of maximum frequency. This provides a physical explanation of the observed energy gap law and isotope effect.

The use of the analogy between (42) and Boltzmann statistical mechanics leads to a simple semi-quantitative description of radiationless processes in aromatic hydrocarbons, but a more accurate approach to the calculation of nonradiative decay rates has also been investigated in other contexts. For the case in which the vibrational modes are harmonic, but need not be parallel or have the same frequencies in the two electronic manifolds s and l, Eq. (40) is mathematically similar to expressions considered by Kubo and Toyazawa in discussions of optical line shapes in solids [50]. In particular, they showed how the double summation in (40) can be expressed as a single definite (Fourier) integral of the form

$$k_{n.r.}(s \to l) = \frac{1}{\hbar^2} \int_{-\infty}^{\infty} f(t) \exp\left(-i\Delta E_k\, t/\hbar\right) dt, \tag{47}$$

where $f(t)$ can be written in explicit closed form. As a large polyatomic molecule is just a finite solid, Lin and Bersohn [8,51] and later Englman and Jortner [49] applied this many-phonon method to the evaluation of nonradiative decay rates. Lin and Bersohn [8,51] considered the complete cumbersome expression (47), but Englman and Jortner [49] noted that in the statistical limit of large ΔE_k, (47) may be evaluated by Laplace's (saddle point approximation) method. Ignoring the effects of the promoting mode and considering only the mode of maximum frequency, they obtained the energy gap law (46) with the pre-exponential factor, ΔE instead of ΔE_k and

$$\gamma = \log\left[\Delta E/d_M\, \hbar\omega_M x_M\right] - 1, \tag{48}$$

where d_M is the degenracy of this mode of maximum frequency. Freed and Jortner generalized the energy gap law to include a proper description of the promoting mode, the C–C skeletal stretches (which have large geometry changes and, hence, Franck-Condon factors in the $T_1 \to S_0$ transition in aromatic hydrocarbons) and the out of plane bending modes [39]. The latter modes suffer a large frequency change in the radiationless transition and might be expected to make significant contributions to the overall Franck-Condon factors. Their treatment involves a generalization of the energy gap law to treat modes with frequency changes [39]. In the case of $T_1 \to S_0$ transitions in aromatic hydrocarbons these out of plane bends are found to have negligible effects on $k_{n.r.}$ in comparison to the C–H and C–C stretches. In the case of frequency changes ΔE_k is replaced by ΔE_k minus the zero point energy change (δE). This correlates well with Siebrand's analysis which gives 5000 cm^{-1} for $\hbar\omega_k + \delta E$ in the aromatic hydrocarbons and 4000 cm^{-1} in the perdeutero compounds (for $T_1 \to S_0$) [22].

It is interesting to note that in the high temperature limit, the pre-exponential factor in the energy gap law becomes $k_B\, T/\hbar$ [39]. This result is not entirely appropriate since temperatures of about 10,000 °K would be required in order to obtain this limit!

11. Effects of Rigid Media on Nonradiative Decay Rates

The energy gap law can also be generalized to discuss the effects of "inert" rigid media upon radiationless transition rates and optical lineshapes of molecules in matrices. This generalization involves the inclusion of both the intra- and intermolecular vibrations in the rate expressions. The simplest type of "inert" medium is one that can only act as a heat bath, i.e., the intermolecular vibrations may have equilibrium displacements or frequency changes which allow them to accept energy upon radiationless transition in the guest molecule. Thus, these intermolecular vibrations may contribute to the overall Franck-Condon factor. It is found theoretically that molecules may exhibit wide phonon broadened optical lines due to large molecule-phonon interactions, and yet these intermolecular vibrations still have negligible effect upon the observed radiationless transition rates. This result is in conformity with experiment and arises because, c.f. (45), the intermolecular phonons have frequencies in the range of ca. $50 - 100$ cm^{-1}. Although they are very numerous, these phonons cannot effectively compete with the higher frequency C–H modes in aromatic hydrocarbons unless the guest-phonon coupling is much larger than that necessary to give wide phonon broadened optical absorption lines. This conclusion has important implications for the measurement of nonradiative decay rates in Shpol'skii matrices which exhibit narrow quasi-line spectra [52]. The narrow lines imply a very weak guest-host interaction in these matrices [53]. Richards and Rice have therefore used the S_2 absorption linewidths of anthracene, naphthalene, coronene and 1, 12 benzperylene dissolved in their appropriate Shpols'kii matrices (n-heptane, n-pentane, n-heptane, and n-hexane respectively) as a direct measure of the $S_2 - S_1$ internal conversion rate [54].

12. Dependence of Nonradiative Decay Rates on Selected Vibronic States

A modified energy gap law has been developed to explicity account for the dependence of the nonradiative decay rate upon the optically selected vibronic level [55,56]. Aside from their intrinsic interest and relevance to the study of photochemistry (see Sec. 13), the dependence of these rates upon the vibronic states is expected to provide a more stringent test of the theory than was provided by the semi-quantitative energy gap law discussed earlier. Spears, Rice and Abramson [23,24] have measured the fluorescence lifetimes and quantum yields from specific vibrational states of electronically excited benzene, perdeutero-benzene and fluorbenzene (S_1 state) at 0.05 and 0.1 torr using monochromatic, (ca. 1.3 Å bandpass) nanosecond flash excitation and time correlated photon counting. [Related experiments have also been performed by Selinger and Ware, Parmenter and Schuyler [25,26], Schlag and von Weyssenhoff, and von Weyssenhoff and Kraus [58].] The present discussion is limited to benzene and perdeuterobenzene.

If the customary assumption is made that there is only a single promoting mode, then in the evaluation of *relative* nonradiative decay rates for different vibronic components of the same electronic state, the *electronic matrix element divides out as a proportionality constant*. The relative rates then depend only upon the details of the two potential surfaces involved, i.e., upon the frequencies and equilibrium positions of the vibrational modes and the energy gap between the two electronic states [23]. Unfortunately, very little is known about the vibrational frequencies, etc. in the T_1 state of benzene. Information is only available for the C−H and C−C totally symmetric vibrations in T_1 [40]. However, these are expected to be the dominant modes in the nonradiative decay. In benzene the dependence of $k_{n.r.}$ on the excitation of the C−C a_{1g} (ν_1) mode in the B_{2u} singlet S_1 has been observed for 0, 1, and 2 quanta, while in perdeutero-benzene it has been observed for 0, 1, 2, and 3 vibrational quanta. Although a large number of other vibronic states have been considered, this progression in the ν_1 mode has been singled out for theoretical study because the greatest amount of data is available for this particular optical mode.

The available energy ΔE_k is partioned between the optical mode and the remainder of the modes which are in their vibrationless state in S_1 ($^1B_{2u}$). The energy partitioning technique, introduced to treat the torsional motion in cis-trans isomerization, [59] can automatically be applied to the problem at hand [55,56]. The optical mode initially has m_a quanta excited while all the other modes are unexcited. In T_1 the optical mode has n_a quanta, while the remainder of the modes take up the rest of the energy

$$\overline{\Delta E_k}(n_a) = \Delta E_k - (n_a - m_a)\, \hbar\omega_a. \tag{49}$$

Here frequency changes have been ignored and ω_a is the optical mode frequency. Thus, the rate $k_{n.r.}(m_a)$ is the sum over all possible n_a of the optical mode Franck-Condon factor $|\langle m_a | n_a \rangle|^2$ times the modified energy gap law for benzene i.e., excluding the optical mode and writing $\Delta E_k \to \overline{\Delta E_k}(n_a)$. For the ν_1 progression in benzene only two values of n_a for each m_a contribute significantly, while in the perdeutero case only three values are necessary [56]. Thus, the summations converge rapidly. The simplest approach considers only the ν_2 C−H a_{1g} vibration in addition to the C−C ν_1 a_{1g} optical mode. Ignoring frequency changes for the moment, the nonradiative decay rates are, apart from terms which are independent of m_a [55,56],

$$k_{n.r.}(m_a) \propto \sum_{n_a} |\langle m_a | n_a \rangle|^2 [\Delta E_k(n_a)]^{-1/2}$$
$$\times \exp\left[-\gamma(n_a)\, \overline{\Delta E_k}(n_a)/\hbar\omega_M\right], \tag{50}$$

where in this case

$$\gamma(n_a) = \ln\left[\overline{\Delta E_k}(n_a)/\hbar\omega_M x_M\right] - 1. \tag{51}$$

The parameters chosen for benzene are $\Delta E = 8200$ cm^{-1} and the promoting mode frequency is taken as $\omega_k = 1500$ cm^{-1}. According to the values given by Burland and Robinson, $\omega_M = 3130$ cm^{-1} and $\omega_a = 923$ cm^{-1}, (the S_1 frequencies) while $x_a \simeq 0.025$ and $x_M \simeq 0.0020$ [40]. For the perdeutero case they give $\omega_M = 2340$ cm^{-1}, $\omega_a = 879$ cm^{-1}, and $\omega_k = 1440$ cm^{-1}. Since x_a and x_M are unknown, x_a is chosen to be the same for the perdeutero and perhydro cases,

Table. *Experimental and calculated relative nonradiative decay rates for the ν_1 progression in the $^1B_{2u}(S_1)$ state of benzene and perdeuterobenzene*

	Benzene		
Relative rate[a]	Experiment[b]	Calc.[c]	(Calc.)'[d]
$k_{n.r.}(1)/k_{n.r.}(0)$	1.22	1.21	1.24
$k_{n.r.}(2)/k_{n.r.}(0)$	1.73	1.44	1.50
	Perdeuterobenzene		
$k_{n.r.}(1)/k_{n.r.}(0)$	1.55 ± .2[e]	1.35	1.40
$k_{n.r.}(2)/k_{n.r.}(0)$	2.61 ± .3	1.76	1.88
$k_{n.r.}(3)/k_{n.r.}(0)$	3.62 ± .5	2.24	2.44

a) Numbers in parenthesis refer to the number of ν_1 quanta excited in the $^1B_{2u}(S_1)$ state.
b) Ref. [24].
c) Parameters given in text.
d) 10% variation in x_a and x_M as given in text.
e) Ref. [57].

while the perdeutero x_M is taken as $\sqrt{2}$ times the perhydro value. (This corresponds to scaling by the ratios of the reduced masses.) The calculated [56] and experimental [24,57] relative rates are presented in Table 1. The theory adequately predicts the isotope effect upon the relative rates: the perdeutero rates increase more sharply upon excitation in ν_1 than the corresponding perdydro rates. This result can simply be understood. As the C–C mode is more excited it becomes more efficient in competing for the available energy ΔE_k. Upon deuteration, ω_M drops more sharply than ω_a and the C–C mode competes more successfully with the C–D mode for ΔE_k. The combination of more efficient nonradiative decay with increasing excitation and the increased inherent effectiveness of the C–C modes in perdeutero compounds leads to the observed effects.

The predicted relative rates do not increase as sharply as the experimental rates. Preliminary calculations imply that frequency changes (of ca. 100 cm^{-1}) in the lower frequency nontotally symmetric modes and/or the promoting mode appear to affect the relatives rates by no more than a few percent. These cannot account therefore for the discrepancy. Included in Table 1 are also the calculated ratios where x_a has been increased by 10% while x_M has been decreased by 10%. Such parameter changes are well within their experimental errors for the perhydro case. These changes tend to provide increased agreement with experiment. In the perdeutero case, the parameters were only "guestimated," and they therefore could be further adjusted in order to obtain improvement with experiment. However, the effects on the relative rates of frequency changes in the C–H and C–C modes should also be investigated, and this work is currently in progress. If these two modes have frequency changes between S_1 and T_1, there is also the possibility that the "normal" modes in T_1 are linear combinations of those in S_1 (only a slight admixture). The neglect of this important mixing as well as anharmonicity effects [40,60] introduce errors into the theoretical calculations. The alternative approach would be first to neglect such mode mixing and anharmonicity and determine a set of triplet vibrational frequencies and oscillator displacements which are consistent with the experimental data. Since these relative rates are sensitive to *frequency changes*, the absolute T_1 frequencies should fairly accurately be obtained. Progressions in the nontotally symmetric modes that have been experimentally observed should be useful in obtaining their properties in T_1 [24,57].

Calculations have shown that the relative rates for the ν_1 progression in benzene and deuterobenzene are extremely sensitive to the shift in this mode's vibrational frequency and that a shift of

$$\nu_1(T_1) - \nu_1(S_1) = + 25 \text{ cm}^{-1}$$

fits the experimental data for both benzene and perdeuterobenzene extremely well [56]. Because of the small isotope effect for this mode, the frequency shift should be the same for both molecules as is predicted by the theory. Similarly, progressions in the ν_6 mode are fit by assigning $\nu_6(T_1)$ frequencies, and the derived $\nu_1(T_1)$ and $\nu_6(T_1)$ frequencies are shown to give theoretical predictions for the relative nonradiative decay rates for the mixed progressions, where the ν_1 and ν_6 modes are both initially excited, which are in excellent agreement with experiment. *These calculations, therefore, exhibit the possibility of assigning vibrational frequencies in an electronic state by measuring radiationless decay rates into that electronic state.* It should be noted that these results are quite insensitive to the choice of the promoting mode frequency [56].

13. Photochemistry

As noted in the introduction photochemical reactions are just one example of a radiationless process. In this case there are rearrangements of chemical bonds and/or breaking of chemical bonds. It is expected that much of the theoretical developments which have provided an understanding of the simpler radiationless processes should also be useful in a proper description of photochemical reactions. The utility of the theories of these simpler processes does not arise only from the fact that photochemical processes often occur simultaneously with the simpler nonradiative decays. It is expected that experimental and theoretical studies of the product yield dependence of photochemical reactions upon the initially selected vibronic state will greatly enhance our understanding of the phenomena (cf., for example, the chemical laser experiments of Pimental and coworkers) [61].

Although the subject of photochemistry is beyond the scope of this review, it is of interest to make a few observations. In considering the vast array of photochemical reactions it is clear that the molecules can "initially" be excited with the electronic excitation localized in some region of the molecule. Subsequently, a bond breaks in the molecule. It is not necessarily the weakest bond that breaks, nor is it necessarily the "excited bond" that breaks. Furthermore the bond that breaks is not necessarily even in close proximity to the region of excitation. What then determines which bond will break? In order to investigate this question qualitatively, a "pin-ball-machine" stochastic model of vibrational relaxation and dissociation was introduced [62]. In this model the photochemical excitation can be introduced into one region of the molecule. The energy "bounces" back and forth between the vibrations of the molecule (the bumpers of the pin-ball-machine), and any number of different bonds are allowed to break with differing intrinsic probabilities (the holes of the pin-ball-machine). The model poses some challenges to and provides a justification for some of the fundamental assumptions of the usual unimolecular rate theories. The model is consistent with the expectation that the product yield distribution be dependent on both the vibrational dynamics and the bond energetics of the molecule. Hence, the above mentioned optical selection studies should be of immense value to our understanding of the phenomena.

14. Conclusions

A unified fully quantum mechanical description of all of the diverse radiationless phenomena has been presented. This provides an understanding of the dissipative and nondissipative aspects associated with radiationless processes in small, large, and intermediate case molecules. The full rate expression is analyzed to provide the observed energy gap law and the associated isotope effects. The theory is generalized to treat nonradiative decay rates in dense media and to evaluate the dependence of these rates on particular selected vibronic states. The relevance of this theory to the study of photochemical processes is also noted.

15. Acknowledgments

This review includes research, both published and unpublished, which was done in collaboration with Professors Joshua Jortner and Stuart Rice, with Dr. William Gelbart, and with Donald Heller. I thank Drs. Ken Spears and Allan Abramson for communicating their results to us prior to publication.

K. F. Freed

16. References

1) Jortner, J., Rice, S. A., Hochstrasser, R. M.: Advanc. Photochem. 7, 149 (1969). –
2) Calvert, J. G., Pitts, J. N., Jr.: Photochemistry, p. 176ff. New York: J. Wiley & Sons
    1966. – Kellog, R. E.: J. Chem. Phys. 44, 411 (1966). – Ware, W. R., Cunningham,
    P. T.: J. Chem. Phys. 44, 4364 (1966).
3) Kellogg, R. E.: J. Chem. Phys. 44, 411 (1966). – Ware, W. R., Cunningham, P. T.: J. Chem.
    Phys. 44, 4364 (1966).
4) Anderson, L. G., Parmenter, C. S.: J. Chem. Phys. 52, 466 (1970).
5) Franck, J., Sponer, H.: Abh. Ges. Wiss. Göttingen 1928, 241. – Kubo, R.: Phys. Rev. 86,
    929 (1952).
6) Robinson, G. W., Frosch, R. P.: J. Chem. Phys. 37, 1962 (1962); 38, 1187 (1963).
7) McCoy, E. F., Ross, I. G.: Australian J. Chem. 15, 573 (1962). – Hunt, G. R., McCoy, E. F.,
    Ross, I. G.: Australian J. Chem. 15, 591 (1962). – Byrne, J. P., McCoy, E. F., Ross, I. G.:
    Australian J. Chem. 18, 1589 (1965).
8) Lin, S. H.: J. Chem. Phys. 44, 3759 (1966).
9) (a) Siebrand, W., Williams, D. F.: J. Chem. Phys. 46, 403 (1967). – Siebrand, W.: J. Chem.
    Phys. 46, 440 (1967);
    (b) Siebrand, W.: The Triplet State (ed. A. B. Zahlan), p. 31. Cambridge University
    Press 1967.
10) Bixon, M., Jortner, J.: J. Chem. Phys. 48, 715 (1968).
11) Kistiakowsky, G. B., Parmenter, C. S.: J. Chem. Phys. 42, 2942 (1965).
12) Anderson, E. M., Kistiakowsky, G. B.: J. Chem. Phys. 48, 4787 (1968). – Parmenter, C. S.,
    White, A. H.: J. Chem. Phys. 50, 1631 (1969).
13) Jortner, J., Berry, R. S.: J. Chem. Phys. 48, 2757 (1968).
14) Chock, D., Jortner, J., Rice, S. A.: J. Chem. Phys. 49, 610 (1968).
15) Bixon, M., Jortner, J.: J. Chem. Phys. 50, 4061 (1969).
16) Freed, K. F.: J. Chem. Phys. 52, 1345 (1970).
17) – Jortner, J.: J. Chem. Phys. 50, 2916 (1969).
18) Bixon, M., Jortner, J., Dothan, Y.: Mol. Phys. 17, 109 (1969).
19) Douglas, A. E.: J. Chem. Phys. 45, 1007 (1967).
20) See also Anderson, L. G., Parmenter, C. S., Poland, H. M., Rau, J. D.: Chem. Phys. Lett.
    8, 232 (1971).
21) Siebrand, W.: J. Chem. Phys. 44, 4055 (1966).
22) – J. Chem. Phys. 47, 2411 (1967).
23) Gelbart, W. M., Spears, K. G., Freed, K. F., Jortner, J., Rice, S. A.: Chem. Phys. Lett.
    6, 345 (1970).
24) Spears, K. G., Rice, S. A.: J. Chem. Phys. 55, 5561 (1971).
25) Ware, W. R., Selinger, B. K., Parmenter, C. S., Schuyler, M. W.: Chem. Phys. Lett. 6,
    342 (1970).
26) Parmenter, C. S., Schuyler, M. W.: Chem. Phys. Lett. 6, 339 (1970).
27) Rice, S. A., Gray, P.: The statistical Mechanics of Simple Liquids. New York: Inter-
    science 1965.
28) Drent, E., Kommandeur, J.: Chem. Phys. Lett. 8, 303 (1971).
29) Rhodes, W.: J. Chem. Phys. 50, 2885 (1969).
30) Messiah, A.: Quantum Mechanics New York: J. Wiley & Sons 1961.
31) Freed, K. F., Gelbart, W. M.: Chem. Phys. Lett. 10, 187 (1971).
32) For instance: Miller, W. H.: Phys. Rev. 152, 70 (1966). – Thesis, Harvard University (1967).
33) Friedland, B., Wing, O., Ash, R.: Principles of Linear Networks. New York:
    McGraw-Hill 1961.

[34] Goldberger, M. L., Watson, K. M.: Collision Theory. New York: J. Wiley & Sons 1964.
[35] Henry, B. R., Kasha, M.: Ann. Rev. Phys. Chem. *19*, 161 (1968).
[36] Rhodes, W., Henry, B. R., Kasha, M.: Proc. Natl. Acad. Sci. (U.S.) *63*, 31 (1969).
[37] Slichter, C. P.: Principles of Magnetic Resonance. New York: Harper & Row 1963.
[38] Wessel, J.: Thesis, The University of Chicago (1970). — Wessel, J., McClure, D.: Proceedings of the Fifth Molecular Crystals Symposium. Philadelphia 1970 (unpublished).
[39] Freed, K. F., Jortner, J.: J. Chem. Phys. *52*, 6272 (1970).
[40] Burland, D, M., Robinson, G. W.: J. Chem. Phys. *51*, 4548 (1969).
[41] — — Proc. Natl. Acad. Sci. U.S. *66*, 257 (1970).
[42] Henry, B. R., Siebrand, W.: J. Chem. Phys. *54*, 1072 (1971).
[43] Lefebvre, R.: Chem. Phys. Lett. *8*, 306 (1971).
[44] Ref. 34, p. 443.
[45] Herzberg, G.: Molecular Spectra and Molecular Structure I, Spectra of Diatomic Molecules, 2nd edit. Princeton: Van Nostrand 1950.
[46] Born, M., Huang, K.: Dynamical Theory of Crystal Lattices. London: Oxford University 1954.
[47] Siebrand, W.: J. Chem. Phys. *54*, 363 (1971).
[48] Tolman, R. C.: The Principles of Statistical Mechanics. London: Oxford University Press 1967. — Hill, T. L.: Statistical Mechanics. New York: McGraw-Hill 1956.
[49] Englman, R., Jortner, J.: Mol. Phys. *18*, 145 (1970).
[50] Kubo, R., Toyozawa, Y.: Prog. Theoret. Phys. *13*, 160 (1955).
[51] Lin, S. H., Bersohn, R.: J. Chem. Phys. *48*, 2732 (1968).
[52] Shpol'skii, E. V.: Usp. Fiz. Nauk. *80*, 255 (1963) [Soviet. Phys. Usp. *6*, 411 (1963)].
[53] Richards, J. L., Rice, S. A.: J. Chem. Phys. *54*, 2014 (1971).
[54] — — Chem. Phys. Lett. *9*, 444 (1971).
[55] Gelbart, W. M.: Thesis, The University of Chicago (1970).
[56] Heller, D. F., Freed, K. F., Gelbart, W. M.: J. Chem. Phys. *56*, 2309 (1972).
[57] Abramson, A.: Thesis, The University of Chicago (1971). — Abramson, A. S., Spears, K. G., Rice, S. A.: J. Chem. Phys. *56*, 2291 (1972).
[58] Weyssenhoff, H. von, Kraus, F.: J. Chem. Phys. *54*, 2387 (1971). — Schlag, E. W., Weyssenhoff, H. von: J. Chem. Phys. *51*, 2508 (1969).
[59] Gelbart, W. M., Freed, K. F., Rice, S. A.: J. Chem. Phys. *52*, 2460 (1970).
[60] Siebrand, W., Williams, D. F.: J. Chem. Phys. *49*, 1860 (1968). — Henry, B. R., Siebrand, W.: J. Chem. Phys. *49*, 5369 (1968).
[61] Berry, M. J., Pimentel, G. C.: J. Chem. Phys. *49*, 5190 (1968); *51*, 2274 (1969); *53*, 3453 (1970).
[62] Gelbart, W. M., Rice, S. A., Freed, K. F.: J. Chem. Phys. *52*, 5718 (1970).

Received April 29, 1971

Fortschritte der chemischen Forschung
Topics in Current Chemistry

Neuere Bände

Band 17
W. Demtröder:
Laser Spectroscopy
With 16 fig. III,95 pages
1971.

Band 18
R. C. Bingham/
P. v. R. Schleyer:
Chemistry of Adamantanes
With 4 fig. III,102 pages
1971

Band 19
L. Maier and G. Zon/
K. Mislow: The Chemistry
of Organophosphorus
Compounds I
With 11 fig. III,94 pages
1971

Band 20
H. J. Bestmann/
R. Zimmermann:
The Chemistry of Organo-
phosphorus Compounds II
With III,147 pages
(In German)
1971

Band 21
L. Eberson/H. Schäfer:
Organic Electrochemistry
With 10 fig. III,182 pages
1971

Band 22
W. Kutzelnigg/G. Del Re/
G. Berthier:
σ and π Electrons
in Organic Compounds
With 8 fig. III, 122 pages
1971·

Band 23
M.J.S. Dewar and
W.B. England/L.S. Salmon/
K. Ruedenberg:
Molecular Orbitals
With 40 fig. and 5 tables
III,123 pages
1971

Band 24
H. Fischer and J.F. Labarre/
F. Crasnier: Electronic
Structure of Organic
Compounds
With 12 fig. III,54 pages
1971

Band 25
J. Manassen, R.L. Banks,
W. Strohmeier,
G.-M.Schwab, F.Steinbach:
Catalysis
With 26 fig. III,154 pages
1972

Band 26
J.L. Margrave/K.G. Sharp/
P.W.Wilson, A. Meller, and
G.D. Christian:
Inorganic and Analytical
Chemistry
With 6 fig. III, 112 pages
1972

Band 27
B. Kratochvil/H.L. Yeager,
V.Gutmann, and S. L.Smith:
Nonaqueous Chemistry
With 46 fig. III, 187 pages
1972

Band 28
G. Häfelinger, J. Tsuji,
L.D. Pettit/D.S. Barnes,
H. Werner:
π Complexes of Transition
Metals
With 11 fig. III, 181 pages
1972

Band 29
P.P.Fietzek/K.Kühn,
H.Clever, H.Krech,
W. Marks, and F.Oehme:
Automation in Analytical
Chemistry
With 32 fig. III, 103 pages
1972

Band 30
A.Weiss, L.Guibé, W.Zeil,
and E.A.C.Lucken:
Nuclear Quadrupole
Resonance
With 23 fig. III, 173 pages
and 4 pages Author index
1972

Springer-Verlag
Berlin
Heidelberg
New York
München · London
Paris · Tokyo · Sydney

All-Valence Electrons S.C.F. Calculations

With 4 figures.
90 pages
1970

Topics in
Current Chemistry
Fortschritte der
chemischen Forschung,
Band 15, Heft 4

**Springer-Verlag
Berlin
Heidelberg
New York**

London München Paris
Sydney Tokyo Wien

The most important
all-valence electron
methods proposed for
S.C.F. calculations of
the properties of large
organic molecules are
discussed. The last five
years have seen the birth
of such methods and
the incredibly fast
development of a
number of more effi-
cient variants designed
to give better agreement
with specific properties.
The trend is undoubted-
ly in favor of the development of an "allpurpose"
method. Some authors
believe this involves the
development of an
NDDO method. Such a
procedure, however,
would require the cal-
culation of a much
larger number of inte-
grals and it would be
difficult to apply it to
large organic molecules
of "chemical interest".

In the opinion of the
present authors, such
calculation would not
improve agreement with
properties found by
experiments because it
would not introduce any
fundamentally new
feature to make up for
the inadequacies of the
present ones. As a mat-
ter of fact, the neglect
of two-center integrals
involving one-center
differential overlap
seems to be a reasonable
hypothesis as shown by
the success of the
M(INDO) methods. On
the other hand, research
workers have usually
confined themselves
to trying to find the
best approximation for
molecular integrals,
while overlooking the
possibility that atomic
orbitals in molecules
might differ widely from
those in the isolated
atoms. (Approx. 80
references)